普通高等教育一流本科专业建设成果教材

化学工业出版社"十四五"普通高等教育规划教材

化工原理实验及虚拟仿真

（双语）

李贤英　主编　　刘建奇　副主编

Chemical Engineering Principle Experiments and Simulation

(Bilingual)

·北京·

内容简介

《化工原理实验及虚拟仿真（双语）》将化工原理实验与计算机仿真、模拟及数据处理结合起来，介绍了实验研究方法、化工原理实验和实验数据处理方法，以及实验室常用测量仪表等内容。本书计算机多媒体仿真、数据模拟采集及处理软件是针对化工原理实验和实验装置，筛选流体、流动、流体传输、传热、蒸馏、吸收、干燥、萃取等单元操作研制开发的。

本书可作为高等院校生物工程、环境工程、环境科学、制药工程、能源、食品工程等专业化工原理实验的教材，也可供有关专业技术人员做参考用书。

图书在版编目（CIP）数据

化工原理实验及虚拟仿真：英、汉/李贤英主编；刘建奇副主编. —北京：化学工业出版社，2024.3

普通高等教育一流本科专业建设成果教材　化学工业出版社"十四五"普通高等教育规划教材

ISBN 978-7-122-44428-8

Ⅰ.①化… Ⅱ.①李… ②刘… Ⅲ.①化工原理-实验-高等学校-教材-英、汉　Ⅳ.①TQ02-33

中国国家版本馆 CIP 数据核字（2023）第 215952 号

责任编辑：满悦芝	文字编辑：曹　敏
责任校对：王　静	装帧设计：张　辉

出版发行：化学工业出版社
（北京市东城区青年湖南街 13 号　邮政编码 100011）
印　　装：三河市双峰印刷装订有限公司
787mm×1092mm　1/16　印张 13¼　字数 334 千字
2024 年 2 月北京第 1 版第 1 次印刷

购书咨询：010-64518888　　　　　售后服务：010-64518899
网　　址：http://www.cip.com.cn

凡购买本书，如有缺损质量问题，本社销售中心负责调换。

定　　价：49.80 元　　　　　　　　　　版权所有　违者必究

前　言

教育领域对接人才全球化的要求使得作为基础教育的本科教育教学国际化成为必然趋势。在高校中实施双语教学有利于促进高等教育国际化进程，提高高等教育的国际竞争能力。采用双语教材，开展双语教学是高等教育改革的趋势，也是质量工程建设的重要内容。采用双语教学有助于学生把握相关背景知识，有助于学生及时跟踪学科发展的最新动态，提高将来进行国际学术交流的能力。对于自然科学和技术等工科类学科，世界各国都拥有共同的术语、定理、公式符号，更适宜进行双语教学。

化工原理是实践性很强的技术基础课，是化工类各专业一门主干课程，属于工程学科范围。同理论教学一样，化工原理实验是整个化工原理教学中的一个重要环节，在培养化工类及相关专业人才中起着举足轻重的作用。我们协同现有的实验教学基础和国外的课程方向，有计划、有系统地使用中文和英文作为教学媒体，使学生在实验课程中学科知识、技术技能以及运用英语能力方面均能顺利和自然地发展。

本教材以东华大学开设的"化工原理实验"课程（上海市精品课程）中的内容和教育部高等学校"化工原理实验"教学大纲为基础，以化工单元操作为主线，参考与借鉴国内外优秀中英文教材，结合东华大学"化工原理实验"多年来在多专业教学中的实践经验编写而成，注重知识的综合应用。东华大学将化工原理实验与计算机仿真、模拟及数据处理相结合，针对化工原理实验和实验装置，对化工单元操作的典型实验研制开发了计算机多媒体仿真、数据模拟采集及处理，并配有相应的多媒体软件——化工原理实验CAI，详情请发送邮件咨询（seanlee@dhu.edu.cn）。本书可作为高等院校化工原理实验教材，也可供生物工程、材料工程、环境工程、环境科学、制药、能源、食品工程及应用化学等相关专业的师生和有关部门的专业技术人员做参考用书。

本书由李贤英任主编，刘建奇任副主编，吴超、王春燕、吴小倩参编。在此感谢有关领导和老师的热情支持与帮助。

由于编写时间仓促，编者的学识和经验有限，对于书中存在的不妥之处，殷切地希望广大读者指正。

编者
2023年12月

Contents

Introduction / 1

Chapter 1　Experimental Research Method of Chemical Engineering Principle / 5
1.1　Dimensional analysis　/ 5
1.2　Mathematical model method　/ 7
1.3　The direct experiment method　/ 9
1.4　Cold-model experiment method　/ 10

Chapter 2　Experiment of Chemical Engineering Principle / 11
Experiment 1　Reynolds experiment　/ 11
Experiment 2　Bernoulli equation experiment　/ 14
Experiment 3　Comprehensive experiment of fluid mechanics　/ 18
　Experiment 3.1　Determination of the pipeline fluid resistance　/ 19
　Experiment 3.2　Calibration of orifice meter and Venturi meter　/ 27
　Experiment 3.3　Determination of the centrifugal pump performance curve　/ 32
Experiment 4　Determination of the characteristic curves of a centrifugal pump　/ 33
Experiment 5　Comprehensive experiment of convective heat transfer　/ 45
Experiment 6　Comprehensive experiment on the measurement of heat transfer coefficient K of heat exchanger　/ 56
Experiment 7　Plate column distillation experiment　/ 59
Experiment 8　Packed column distillation experiment　/ 72
Experiment 9　Determination of fluid mechanics of plate column　/ 75
Experiment 10　Absorption-desorption experiment　/ 81
Experiment 11　Drying experiment　/ 89
Experiment 12　Fluidized bed drying experiment　/ 96
Experiment 13　Extraction experiment　/ 103
Experiment 14　Pervaporation membrane experiment　/ 107
Experiment 15　Ultrafiltration membrane separation experiment　/ 111

Chapter 3　Methods of Processing Experimental Data / 113
3.1　Significant digits and operation rules　/ 113
3.2　Error analysis of experimental data　/ 114
3.3　Experimental data processing　/ 116

3.4　Representation of experimental data with mathematical equation　/ 123

Chapter 4　Common Measuring Instrument in Laboratory　/ 132
4.1　Pressure measurement　/ 132
4.2　Metering of fluids　/ 139
4.3　Temperature measurement　/ 153

Appendix　/ 166
Appendix 1　SXK-2 type high precision flow integrator　/ 166
Appendix 2　ZW5433 three-phase digital meter　/ 168
Appendix 3　Instructions for hydrostatic balance　/ 170
 Appendix 3.1　Specific gravity table of ethanol solution　/ 172
 Appendix 3.2　Density table of water　/ 175
Appendix 4　Abbe refractometer　/ 175
 Appendix 4.1　Relationship between refractive index of ethanol-propanol and solution concentration　/ 178
 Appendix 4.2　Relationship between composition and refractive index of n-heptane-methyl cyclohexane system　/ 178
Appendix 5　Gas-liquid equilibrium data　/ 179
 Appendix 5.1　Gas-liquid equilibrium data of ethanol-aqueous solution at atmospheric pressure (P= 760mmHg)　/ 179
 Appendix 5.2　Gas-liquid equilibrium data of ethanol-propanol at atmospheric pressure (molar fraction)　/ 179
 Appendix 5.3　Gas-liquid equilibrium data of n-heptane methyl cyclohexane at atmospheric pressure　/ 180
Appendix 6　Characteristic parameters of the four packings　/ 180
Appendix 7　Oxygen dissolving instrument　/ 180
Appendix 8　752-model spectrophotometer　/ 182
Appendix 9　Principles of automatic acquisition of computer data and automatic control　/ 184
Appendix 10　Installation and operation instructions of "Chemical Engineering Principle Experiment CAI"　/ 185
Appendix 11　Data recording and collation table of unit operation experiment　/ 195
 Appendix 11.1　Data recording and collation table of Reynolds experiment　/ 195
 Appendix 11.2　Data recording and collation table of Bernoulli equation experiment　/ 195
 Appendix 11.3　Data recording and collation table of the pipeline fluid resistance determination　/ 196
 Appendix 11.4　Data recording and collation table of calibration experiment of orifice and Venturi flowmeter　/ 197
 Appendix 11.5　Data recording and collation table of determination of centrifugal pump characteristic curve　/ 198
 Appendix 11.6　Data recording and collation table of comprehensive heat transfer experiment　/ 199
 Appendix 11.7　Data recording and collation table of comprehensive experiment of heat transfer coefficient K measurement　/ 200

Appendix 11. 8　Data recording and collation table of plate column distillation experiment　/ 201
Appendix 11. 9　Data recording and collation table of packed column distillation experiment　/ 201
Appendix 11. 10　Data recording and collation table of determination of hydromechanical properties of plate column　/ 202
Appendix 11. 11　Data recording and collation table of absorption-desorption experiment　/ 202
Appendix 11. 12　Data recording and collation table of drying experiment　/ 204
Appendix 11. 13　Data recording and collation table of the fluidized bed drying experiment　/ 205
Appendix 11. 14　Data recording and collation table of extraction experiment　/ 206
Appendix 11. 15　Data recording and collation table of pervaporation membrane experiment　/ 207
Appendix 11. 16　Data recording and collation table of ultrafiltration membrane separation experiment　/ 208

References　/ 209

Introduction

绪 论

(1) The objective of chemical engineering principle experiment（化工原理实验目的）

As an engineering experimental course, the experiment of chemical engineering principle is an important part of chemical engineering teaching, which is based on the principle of chemical engineering and complementary to it. Different from basic experiment, engineering experiment faces with complex practical engineering problems. Aiming at the different objects, the experimental research methods inevitably differs. The experiment of chemical engineering principle not only deepens the students' understanding of the basic principle of chemical engineering, but also plays an important role in the basic training of experimental research methods and experimental skills. In the experimental operations, students can improve their ability of acute insight to observe experimental phenomena, the ability to use various experimental methods to obtain valid experimental data, and the ability to analyze and summarize experimental data and phenomena. Students can comprehensively utilize the theoretical knowledge to draw conclusions and put forward their own opinions based on experimental data and phenomena, finally, achieve the purpose of strengthening innovation consciousness and improving the ability to analyze and solve practical problems.

Therefore, the following objectives are expected to achieve through experiments:

① Verify the basic theory of chemical process, use the theory to analyze the experimental process, so as to further understand and consolidate the theoretical knowledge.

② Familiar with the process, structure and operation of experimental equipment, as well as the use of common instruments in chemical industry.

③ Master basic testing techniques of chemical data, such as measuring methods of operating parameters (pressure, flow, temperature, etc.), equipment characteristic parameters (resistance coefficient, heat transfer coefficient, etc.) and characteristic curves.

④ Enhance the engineering concept, master the experimental research method, cultivate the ability to design and organize experiments.

⑤ Improve the ability of data processing and problem analysis. Complete the experimental reports.

(2) Characteristics of computer simulation, data simulation acquisition and processing for chemical principle experiment（化工原理实验计算机仿真、数据模拟采集及处理的特点）

To adapt to the development of modern educational technology as well as the trend of scientific intersection and synthesis, the combination of experimental teaching content, experimental process, data processing and computer technology of chemical engineering principle as a new teaching method has changed the traditional teaching mode and provided a human-computer interactive environment for students. Chemical engineering principle experimental CAI is a computer simulation experimental system, which refers to the use of computer graphics technology to simulate the chemical engineering principle experimental device on the monitor screen, and to simulate the experimental operation process through computer input device (mouse or keyboard). By means of chemical mathematical model and computer numerical calculation, the variation of chemical experimental parameters in the process of operation as well as the actual experimental data processing and the result diagram is simulated. The courseware has the following characteristics:

① The courseware interface is friendly, simple, visual and expressive, which realizes interface control and multimedia information transmission.

② The courseware includes the demonstration of relevant experiment materials, and can be used as experiment preview and practice to achieve more effective experiment results.

③ The courseware can simulate the abnormal operation and the phenomenon that is not easy to observe in the experiment.

④ The courseware has the function of providing basic data and data processing of chemical industry, which can be used as a tool to check students' experimental results.

(3) Requirements for chemical principle experiments (化工原理实验的要求)

The chemical principle experiment should include experimental preview, experimental operation, determination, data recording, data collation, experiment report compilation and other steps (化工原理实验应包括实验预习、实验操作、测定、记录和数据整理、实验报告编写等步骤). In order to complete the experiment smoothly and make the qualified experiment report, the following instructions and requirements are put forward for each step in the experiment process.

① Preparation for the experiments

a. Read the experimental instruction carefully, and make clear the content and requirements of the experiments.

b. According to the specific task of this experiment, study the theoretical basis of the experiment and the specific operations of the experiment, analyze what kind of data should be measured, and estimate the variation rule of these data.

c. Basing on the experiment content, carefully understand the experimental process, the structure of the main equipment, the installation of the instrument, measurement principles and methods at the laboratory site. According to the experiment task and site investigation, draw up the experimental plan and operation steps.

d. Combining with the experimental multimedia simulation software, carry on the computer simulation experiment and the multimedia demonstration of related experimental materials.

② Experimental operations Chemical principle experiments are generally carried out in groups of 3-4 students. The members of the experimental group are required to perform their respective duties (including operation, reading and recording data, etc.) during the experimental operation, and rotate the job at the appropriate time to complete the experiment by teamwork.

a. Check the experimental equipment before starting（实验设备启动前必须检查）

(a) Whether the opening and closing status of each valve on the equipment and pipeline meets the requirements of the process（设备、管道上各个阀门的开、关状态是否符合流程要求）.

(b) For running equipment such as pumps, check whether it can rotate normally before starting the equipment（泵等转动的设备，启动前先盘车检查能否正常转动，才可启动设备）.

(c) Master the correct use of the instrument（掌握仪表的正确使用方法）.

b. Students should concentrate on operating, record the experimental data, observe the experimental phenomenon, and deal with the problems in time or report to the experiment instructor.

c. At the end of the experiment operation, the valves of the gas source, water source, heat source and test instrument should be turned off successively, the power supply of the main equipment should be cut off, and the opening or closing position of each valve should be adjusted.

d. Input the experimental data into the computer to check whether the experiment is correct, and redo the experiment if there is any error（将读取的实验数据输入计算机处理，检验实验是否正确，如果有错即重做实验）.

③ Measurement, data record and data collation（实验测定、记录和数据整理）

a. Experimental data　　All necessary parameters affecting the experimental results should be measured. It includes atmospheric conditions, equipment dimensions, physical properties and operating data. For data that can be derived from a specific data or a manual, it is not necessary to determine directly. For example, the density, viscosity, specific heat and other physical properties of water can be found out as long as the water temperature is measured.

b. Read and record experimental data

(a) According to the purpose of the experiment, the data record form should be made before the experiment, and the title, symbol and unit of each physical quantity should be noted in the table.

(b) In the experiment, the data should not be read until the phenomenon is stable. If the condition changes, it is also necessary to read the data after a certain time of stability in order to exclude the inaccurate readings by the hysteresis phenomenon of the instrument.

(c) The recorded data should be reviewed immediately to avoid misreading or clerical error.

(d) Data records must reflect the accuracy of the instrument. Generally, one digit below the minimum scale of the instrument should be reserved.

(e) The abnormal conditions in the experiment or obvious errors in the data should be mentioned in the note column.

c. Collation of experimental data

(a) The original recorded data can only be collected but never modified（原始记录数据只可进行整理，绝不可修改）. The incorrect data caused by gross error may not be included in the results.

(b) Several fluctuating data at the same experimental point can be averaged and then collated.

(c) Use a tabular method to collate data for clear comparison. A calculation example should be attached to the table to illustrate the relationship between the items（在表格之后应附计算示例，以说明各项之间的关系）.

(d) In the operation, applied constant (i. e. , transformation factor) is suggested as far as possible, as shown as tabular method in experimental data processing method for details.

(e) The experimental results can be expressed in the form of lists, curves, graphs or equations.

④ Compile the reports（实验报告的编写） The experimental report must be simple and concise with complete data and correct conclusions. Formulas or curves are obtained after discussion and analysis. The experimental conditions should be clearly indicated in the diagram. The format of the report should generally include the following：

a. Experiment title.

b. The names of the author and the members of the same experimental group are usually written in the upper right corner of the first page of the report.

c. Experimental purpose.

d. Experimental principle.

e. Flow chart and description of experimental equipment (should include flow diagram, types and specifications of main equipment).

f. Experimental data (Record of experimental data and experimental data collation).

g. Calculation example：explain the source of the quoted data, write out the derivation process of the simplified formula, and listed the complete calculation process of a certain set of data as calculation example（计算示例，其中引用的数据要说明来源，简化公式要写出导出过程，要列出某一组数据的完整计算过程作为计算示例）.

h. Experimental results and discussion：according to the experimental task, draw the conclusion by graphic method, empirical formula or list method, and note all the experimental conditions. Further evaluate the experimental results, analyze the errors and causes. Necessary discussion and suggestions for experimental methods and equipment can also be included in this part.

Chapter 1　Experimental Research Method of Chemical Engineering Principle

第 1 章　化工原理实验研究方法

Based on summing-up of experience and long-term experimental research, the progressive research method in experiment of chemical engineering principles mainly include dimensional analysis, mathematical model, direct experiment and cold mold experiment method.

1.1　Dimensional analysis
（量纲分析法）

Many important engineering problems cannot be solved entirely by theoretical or mathematical methods. These problems are especially common in fluid-flow and heat transfer operation. For example, it involves the geometrical conditions of equipment, fluid flow patterns, fluid physical property changes and other factors. One way attacking these problems is to experiment empirically. To study the law of multivariate influence process, the common approach is systematically varying one variable while keeping all others constant （为了研究多变量影响过程的规律，常用的方法是系统地改变一个变量，而保持所有其他变量不变）. However, if the number of variables is m and the condition of each variable changes for n times, the experiment number required is n^m. Following this approach, the experimental workload is too large to implement. There is an empirical method to drastically simplify the task and obtain a universal relationship, which is *dimensional analysis*（量纲分析法）.

The dimensional analysis is based on the fact that if a theoretical equation does exist among the variables affecting a physical process, that equation must be dimensional consistency（量纲一致性）. The Buckingham π theorem is a key theorem in dimensional analysis. The theorem loosely states that if we have a physically meaningful equation involving a certain number, n, of physical variables, and these variables are expressible in terms of k independent fundamental physical quantities, then the original expression is equivalent to an equation involving a set of $p = n - k$ dimensionless variables constructed from the original variables. Because of this requirement, it is possible to group the many factors into a smaller number of dimensionless groups of variables by experiments（可将多

变量函数整理为量纲 1 数群函数，然后通过实验归纳成准数关系式），so as to reduce the experimental workload. Most importantly, it provides a method for engineering design calculation from the given variables, even if the form of the equation is still unknown.

For example, the correlation formula of resistance and friction coefficient of fluid in the pipe is obtained by dimensional analysis method and experiment（流体在管内流动的阻力和摩擦系数关联式是利用量纲分析法和实验而得出的）. The experimental results show that the factors affecting fluid resistance in turbulent flow are the pipe diameter d through which the fluid flows, the pipe length L, the average velocity u, the physical properties of density ρ, the viscosity μ and the roughness ε of the pipe wall. If a theoretical equation for this problem exists, it can be written in the general form:

$$\Delta p = f(d, L, u, \rho, \mu, \varepsilon) \tag{1-1}$$

If the Eq. (1-1) is a valid relationship, all terms in the function must have the same dimensions as the left-hand side of the equation, Δp. Let the dimensions be shown by the use of square brackets. Then any term in the function must conform to the dimensional formula:

$$[\Delta p] = K [d]^a [L]^b [u]^c [\rho]^e [\mu]^f [\varepsilon]^g \tag{1-2}$$

In the formula, the constant K and the exponents a, b, c, e, f, g are undetermined values. There are only three fundamental physical units in this equation: mass $[M]$, time $[\theta]$, and length $[L]$:

$$[p] = ML^{-1}\theta^{-2} \qquad\qquad [\rho] = ML^{-3}$$
$$[d] = L \qquad\qquad [\mu] = ML^{-1}\theta^{-1}$$
$$[u] = L\theta^{-1} \qquad\qquad [\varepsilon] = L$$

Substituting the dimensions into Eq. (1-2) gives

$$ML^{-1}\theta^{-2} = K [L]^a [L]^b [L\theta^{-1}]^c [ML^{-3}]^e [ML^{-1}\theta^{-1}]^f [L]^g \tag{1-3}$$

Namely:
$$ML^{-1}\theta^{-2} = K [M]^{e+f} [L]^{a+b+c-3e-f+g} [\theta]^{-c-f} \tag{1-4}$$

Since Eq. (1-4) is assumed to be dimensional consistency, the exponents of the individual primary units on the left-hand side of the equation must equal those on the right-hand side. This gives the following set of equations:

Exponents of $[M]$: $e + f = 1$

Exponents of $[L]$: $a + b + c - 3e - f + g = -1$

Exponents of $[\theta]$: $-c - f = -2$

Here are six variables but only three equations. Three of the unknowns may be found in terms of the remaining three, as follows:

$$a = -b - f - g$$
$$c = 2 - f$$
$$e = 1 - f$$

By substituting a, c and e into Eq. (1-2):

$$\Delta p = K d^{-b-f-g} L^b u^{2-f} \rho^{1-f} \mu^f \varepsilon^g \tag{1-5}$$

Collecting all factors having same exponents into one group, gives:

$$\frac{\Delta p}{\rho u^2} = K\left(\frac{L}{d}\right)^b \left(\frac{du\rho}{\mu}\right)^{-f} \left(\frac{\varepsilon}{d}\right)^g \tag{1-6}$$

The above formula consists of only four dimensionless groups: L/d reflects length-to-diameter ratio of the pipe, which relates to the geometric size of the pipe. $du\rho/\mu$ is Reynolds number reflecting the flow characteristics of the fluid. ε/d known as relative roughness is the ratio of absolute roughness to pipe diameter, which relates to the material of the pipe. $\Delta p/\rho u^2$ represents the ratio of pressure to inertia force, known as the Euler number, usually expressed in Eu.

From the π theorem, the total number of physical quantities affecting the process is seven and the basic dimension is three, so the above four dimensionless quasi numbers can be obtained.

In addition, Eq. (1-6) can be sorted out as:

$$h_f = \frac{\Delta p}{\rho} = 2K\left(\frac{L}{d}\right)^b \left(\frac{du\rho}{\mu}\right)^{-f} \left(\frac{\varepsilon}{d}\right)^g \frac{u^2}{2} \tag{1-7}$$

Fanning equation:

$$h_f = \lambda \frac{L}{d} \frac{u^2}{2} \tag{1-8}$$

Comparing the above two formulas gives:

$$\lambda = 2K\left(\frac{du\rho}{\mu}\right)^{-f} \left(\frac{\varepsilon}{d}\right)^g \tag{1-9}$$

or:

$$\lambda = \varphi\left(Re, \frac{\varepsilon}{d}\right)$$

From the above analysis, the dimensional analysis can simplify a complex multi-variable fluid resistance calculation in the pipe into the experimental studying by determining the specific function relationship of friction coefficient λ. In addition, when making dimensional analysis, it is necessary to make a thorough understanding of process studied and choose the important relevant variables. If some variables are left out or some of those chosen are not needed, it will lead to incorrect results or even wrong conclusions.

1.2 Mathematical model method（数学模型法）

Established on the basis of in-depth research and full understanding of the inherent laws of the process, mathematical model can be thought of an imaginary, simplified version of the physical model of an actual complex process, and expressed by mathematical equations （数学模型法是建立在对过程的内在规律做深入的研究和充分的认识基础上对复杂的问题高度概括得出简化而又近似实际过程的物理模型并用数学方程表示的方法）. In the following example, the mathematical model of mass transfer rate is introduced by calculating the packing height according to the absorption rate equation.

Liquid and gas continuously contact with each other at any given section through the column. The composition of the liquid (X) and gas (Y) change continuously along the packing height, then the mass transfer driving force ΔY (or ΔX) and mass transfer rate change accordingly. Therefore, the analysis must start from the solute absorption process of a differential packing in height dz. As shown in Fig. 1-1, assume the mass transfer area (the interfacial area between phases L and V) in the differential packing dz is dA:

$$dA = \Omega a \, dz \qquad (1\text{-}10)$$

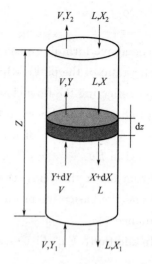

Fig. 1-1 Material balance of differential packing layer

图 1-1 微元填料层的物料衡算

Where, dA——mass transfer area in the differential packing dz, m^3;

Ω——sectional area of the column, m^2;

a——effective contact area per unit volume packing (known as effective specific surface area), m^2/m^3.

Based on the material balance:

After gas passes through dz, the amount of solute being absorbed is:

$$dG = V \, dY \qquad (1\text{-}11)$$

After liquid passes through dz, the amount of solute being absorbed is:

$$dG = L \, dX \qquad (1\text{-}12)$$

Based on the gas or liquid phases, the rate of absorption per unit volume can be expressed as:

$$dG = K_Y (Y - Y^*) \, dA \qquad (1\text{-}13)$$

or

$$dG = K_X (X^* - X) \, dA \qquad (1\text{-}14)$$

Where, K_Y——overall absorption coefficient of gas phase with $(Y - Y^*)$ as the overall driving force, $kmol/(m^2 \cdot s)$;

K_X——overall absorption coefficient of liquid phase with $(X^* - X)$ as the overall driving force, $kmol/(m^2 \cdot s)$;

Y^*——the composition of the gas phase in equilibrium with the liquid phase composition X;

X^*——the composition of the liquid phase in equilibrium with the gas phase composition Y.

Correlating equations of (1-10), (1-11) and (1-13) gives:

$$dZ = \frac{V}{K_Y a \Omega} \frac{dY}{Y - Y^*} \qquad (1\text{-}15)$$

Similarly, correlating equations of (1-10), (1-12) and (1-14) gives:

$$dZ = \frac{L}{K_X a \Omega} \frac{dX}{X^* - X} \qquad (1\text{-}16)$$

a in all these coefficients is the interfacial area per unit volume of the packed column or other device, and it is smaller than the solid surface area (called specific surface area) in the packing per unit volume. The value of a is not only related to the packing form, size and packing condition, but also affected by the fluid physical properties and flow pattern. It is hard to measure or to predict a, but in most cases the product of K_Y and a (or K_X and a) is regarded as a whole, which is called volumetric mass transfer coefficient. In the experimental study, $K_Y a$ (or $K_X a$) is measured together, the unit is $kmol/(m^3 \cdot s)$. Its physical significance is the mass of solute being absorbed in the packing per unit driving force, per unit time and per unit volume.

According to the separation requirements, the column height Z can be obtained by integrating Eq. (1-15):

$$Z = \int_{Y_2}^{Y_1} \frac{V}{K_Y a \Omega} \frac{dY}{Y - Y^*} \tag{1-17}$$

Ω and V are constants in the upper formula. For dilute gases, the change in molar flow rate for both gas and liquid including solute is neglected. The physical properties of gas and liquid phase also change little, so volumetric mass transfer coefficient $K_Y a$ on each section does not change much, which can be regarded as a constant independent of the height of the column. The equation for column height can be written as follows:

$$Z = \frac{V}{K_Y a \Omega} \int_{Y_2}^{Y_1} \frac{dY}{Y - Y^*} \tag{1-18}$$

In similar to Eq. (1-18), the Eq. (1-16) is rearranged for integration:

$$Z = \frac{L}{K_X a \Omega} \int_{X_2}^{X_1} \frac{dX}{X^* - X} \tag{1-19}$$

From Eq. (1-18),

$$K_Y a = \frac{V}{Z \Omega} \int_{Y_2}^{Y_1} \frac{dY}{Y - Y^*} \tag{1-20}$$

The above formula is the mathematical model for the determination of the overall volumetric absorption coefficient of gas phase by absorption experiment.

1.3 The direct experiment method
（直接实验法）

The direct experiment method is to conduct direct experiment on the studied object to obtain the relevant parameters and rules（直接实验法即对被研究的对象进行直接的实验以获取其相关的参数及规律）. The results obtained by direct experiment for specific engineering problems are more reliable. It is a direct and effective method to solve engineering problems, which other research methods cannot solve. However, this method also has great limitations. The conclusion obtained is merely the rule of the relationship between individual parameters, which cannot reflect the essence of the object. Thus, these experimental results can only apply to specific experimental conditions and equipment, or generalize to the phenomenon of identical experimental conditions. In addition, the experiment is heavy workload, time-consuming, and sometimes high investment.

1.4 Cold-model experiment method（冷模实验法）

The cold-model experiment is to analyze and speculate the actual process by simulating the experimental results, and mainly used to simulate the flow state, transfer process and other physical processes（冷模实验主要用于流动状态、传递过程等物理过程模拟研究，通过模拟实验结果去分析、推测实际过程）. For example, the experimental study of gas-liquid mass transfer using air, water and tracers provides parameters for the design and modification of gas-liquid mass transfer equipment; the experimental study of fluidized bed reactor using air and sand provides the basis for the design of fluidized bed reactor. Therefore, using air, water, sand and other simulated materials instead of real materials to study the influence of various engineering factors on the process in the experimental equipment with similar structure and size of industrial equipment, is called "cold-model experiment". The advantages of cold-mode experiment are summarized as follows：

① The results of cold-model experiment can be extended to other actual fluids. The experimental results of small-size experimental equipment can be promoted to large industrial equipment, so that the experiment can be "from one to another" in terms of material types, and "from small to large" in terms of equipment size.

② Its intuitiveness and economy. The relationship between various physical quantities can be obtained through fewer experiments combined with mathematical model method or factor analysis method, which greatly reduces the experimental workload.

③ Analogical experiments that are inconvenient or impossible under real conditions can be implemented to reduce the risk of experiments.

It is worth pointing out that the results of cold-mode experiments must combine with chemical reactions and other characteristics. After correction, it can be used in the design and development of industrial processes.

Chapter 2　Experiment of Chemical Engineering Principle

第 2 章　化工原理实验

Experiment 1　Reynolds experiment
实验 1　雷诺实验

I. Experiment purpose

① Establish the perceptual understanding of "laminar flow and turbulent flow".
② Observe the relationship between Reynolds number（雷诺数）and fluid flow type.
③ Observe the velocity distribution of laminar flow in circular tube.

II. Experimental principle

In laminar flow（层流）, the fluid particles move in a straight line, that is, the fluid flows in layers without macroscopic mixing with the surrounding fluid. In turbulent flow（湍流）, the fluid particles pulsate randomly in all directions, while the fluid generally flows along the direction of the pipeline.

Reynolds number is the criterion to judge the actual flow type（雷诺数是判断实际流动类型的准数）. When the fluid flows in a circular tube, the Reynolds number can be expressed as:

$$Re = \frac{du\rho}{\mu} \tag{2-1}$$

Where, d—diameter of tube;
　　　　u—average velocity of fluid;
　　　　μ—viscosity of fluid;
　　　　ρ—density of fluid.

In a pipe or tube, flow is always laminar at Reynolds numbers below 2000, while the flow is turbulent at Reynolds numbers above about 4000. Between 2000 and 4000, there is a transition region, where the flow may be either laminar or turbulent, depending upon conditions at the entrance of the tube and on the distance from the entrance. Re value at the end of laminar flow is called the critical Reynolds number（临界雷诺数）, or the critical value for short.

For a fluid in a particular tube at a given temperature, the Reynolds number only relates to the velocity of flow. This experiment is to change the velocity of water in the tube and observe the change of fluid flow pattern under different Reynolds number.

Theoretical analysis and experimental results show that the velocity of laminar flow is parabola distribution along the pipe diameter（层流时的速度沿管径按抛物线的规律分布）. The velocity is the fastest in the center of the tube, and the velocity decreases as it approaches the wall. In turbulent flow, the velocity distribution is no longer strictly parabolic because the fluid particles are strongly separated and mixed. The stronger the turbulence, the wider and flatter the top region of the velocity distribution curve. However, even in turbulent flow, the fluid near the wall of the tube still flows in laminar flow, which is laminar sublayer（层流内层或层流底层）. It is extremely thin, but in the transfer of heat and mass in the fluid, the resistance is much greater than the turbulent body of the fluid.

III. Experimental device and process（实验装置及流程）

(1) Schematic diagram of experimental equipment and flow chart（实验装置示意图及流程）

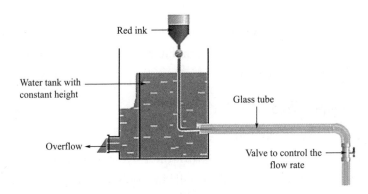

Fig. 2-1　Reynolds experiment—Device schematic diagram and flow chart
图 2-1　雷诺实验——装置示意图及流程

The experimental device is shown in Fig. 2-1, water in the tank is supplied by the water pipe. During the experiment, the water enters the glass tube from the water tank（observe the flow pattern of the fluid and the velocity distribution in the horizontal glass tube during laminar flow）. The water flow is controlled by the outlet valve. The water tank is equipped with an inlet steady flow device and an overflow pipe to maintain a stable and constant liquid level. The excess water is discharged from the overflow pipe into the sewer.

(2) Experimental simulation interface（Fig. 2-2）（雷诺实验——仿真界面）

IV. Experimental steps（实验步骤）

(1) Experimental operations

① Reynolds experiment（雷诺实验）

a. Open the inlet valve（进水阀）to fill the high water tank with tap water；

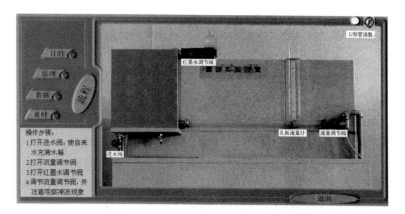

Fig. 2-2　Reynolds experiment—Simulation interface
图 2-2　雷诺实验——仿真界面

b. After overflow（溢流），open the flow control valve；

c. Slowly open the red ink regulating valve；

d. Adjust the flow control valve（流量调节阀）and observe the laminar flow phenomenon；

e. Gradually open the flow control valve, and observe the phenomenon of transition flow；

f. Further increase the opening of the flow control valve, and observe the turbulence phenomenon；

g. Measure the flow rate of the fluid by the orifice flowmeter and calculate the Reynolds number；

h. Close the red ink regulating valve, and then close the water inlet valve. When the red in the glass tube disappears, close the flow control valve and end the experiment.

② Demonstration of the velocity distribution curve of the fluid in a circular tube（流体在圆管内的速度分布曲线的演示）

a. Open the inlet valve to fill the water tank with tap water；

b. Open the red ink regulating valve and deposit a certain amount of red ink in the glass duct；

c. Quickly open the flow control valve to a large opening, and observe the velocity distribution curve of the fluid in the circular glass tube；

d. Close the red ink regulating valve, and then close the water inlet valve. Until the red in the glass tube disappears, and then close the flow control valve.

（2）Operation notes（注意事项）

① When opening the water inlet valve of the water tank, attention should be paid to the control of the influent quantity（在开启水箱的进水阀时，应注意控制进水量）, so that it is slightly larger than the water consumption（at this point, a little water overflow is observed in the overflow pipe）. If the influent is too large, the overflow flow will be too much. Under the interference of a large amount of overflow, the liquid level will fluctuate seriously and affect the experimental results.

② The amount of red ink should not be too large to avoid waste and affect the experimental results.

③ If the equipment is long outage, discharge the water stored in each part of the experimental device.

V. Record and collation of experimental data (Fig. 2-3)

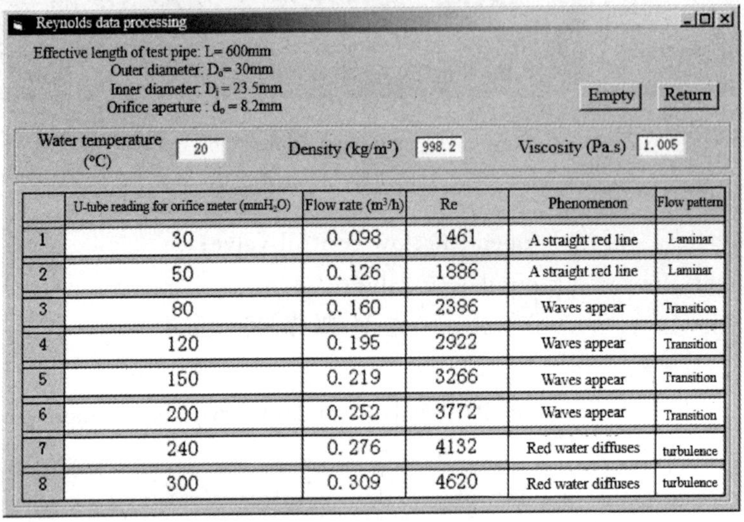

	U-tube reading for orifice meter (mmH$_2$O)	Flow rate (m^3/h)	Re	Phenomenon	Flow pattern
1	30	0.098	1461	A straight red line	Laminar
2	50	0.126	1886	A straight red line	Laminar
3	80	0.160	2386	Waves appear	Transition
4	120	0.195	2922	Waves appear	Transition
5	150	0.219	3266	Waves appear	Transition
6	200	0.252	3772	Waves appear	Transition
7	240	0.276	4132	Red water diffuses	turbulence
8	300	0.309	4620	Red water diffuses	turbulence

Effective length of test pipe: L= 600mm
Outer diameter: D$_o$= 30mm
Inner diameter: D$_i$= 23.5mm
Orifice aperture: d$_o$= 8.2mm
Water temperature (°C): 20
Density (kg/m^3): 998.2
Viscosity (Pa.s): 1.005

Fig. 2-3 Reynolds experiment—Record and organize experimental data

图 2-3 雷诺实验——数据记录及整理

VI. Questions

① What is the relationship between the flow pattern and the value of the Reynolds number?

② What are the factors affecting the fluid flow pattern?

③ Observe the shape of the top of the red ink when the liquid velocity is distributed in the tube. What does this shape mean?

④ If the pipe is opaque, the flow pattern in the pipe cannot be determined by direct observation. What method do you think is available to determine the flow pattern in the pipe?

Experiment 2 Bernoulli equation experiment
实验 2 伯努利方程实验

I. Experiment purpose

① Deepen the understanding of various energy or pressure heads in the flowing fluid and their mutual conversion concepts, and master Bernoulli equation.

② Observe the change rule of flow velocity to understand the continuity equation of fluid flow.

③ Observe the mutual conversion rule of various pressure heads.

II. Experimental principle

During the steady flow of incompressible fluid in the pipeline, due to the change of pipeline conditions (such as position, diameter, distance), various mechanical energy will self-transform（各种机械能之间自行转化）. The relationship can be described by the energy balance of Bernoulli equation（伯努利方程）in the process of flow:

$$z_1 g + \frac{u_1^2}{2} + \frac{p_1}{\rho} = z_2 g + \frac{u_2^2}{2} + \frac{p_2}{\rho} + \sum h_f \tag{2-2}$$

For an ideal non-viscous fluid（无黏性的理想流体）, there is no friction or collision（无摩擦和碰撞）between the fluid particles, and no loss of mechanical energy. That is, $\sum h_f = 0$. In this case, the sum of the mechanical energies is the same although each of mechanical energy at any two cross-sections of the pipeline is not necessarily equal. For actual fluids, part of mechanical energy consumes in the flow process due to the internal friction generated by viscosity, which converted into heat energy and cannot be recovered. Thus, in terms of the actual fluid, the sum of mechanical energies at the two cross-sections is not always equal. The difference between the two is the mechanical energy loss of the fluid. Therefore, the lost mechanical energy must be added to the downstream section when calculating the mechanical energy.

All of the above mechanical energies can be expressed by the liquid column height in the piezometer tube, which is called "pressure head（压头）". The potential head（势能压头或位压头）of the liquid at the measuring tube is determined by the geometric height of the pressure tap（测量管处液体的位压头则由测量管的几何高度决定）. The kinetic energy is called dynamic head (or velocity head)（动压头或速度头）. The pressure energy is called the static head (or pressure head)（静压头或压强压头）. The static pressure measuring tube is perpendicular to the flow direction, and the height of liquid column in the measuring tube from the measuring hole is the static head, which reflects the magnitude of the hydrostatic pressure at the pressure measuring point. The mechanical energy loss is called the loss head (or friction head)（损失压头或摩擦压头）.

If the pressure tap is facing the direction of water flow, the height of the measured liquid column is the stamping head（冲压头）, which is the sum of the static head and the dynamic head（冲压头即为静压头和动压头之和）. Between any two cross-sections, the difference between the sum of the potential head, the dynamic head and the static head is the loss head, which represents the mechanical energy consumption of the fluid flowing through the two cross-sections.

III. Experimental device and process

(1) Schematic diagram of experimental equipment and flow chart

The experimental device consists of test glass tube, measuring tube, stainless steel centrifugal pump, high and low water tank. The glass flow tube is made into four sections of different diameter and height, as shown in Fig. 2-4. The diameter of the thin pipe is 14mm, that of the thick pipe is 28mm, and the height difference between the center horizontal position of high pipe and low pipe is 90mm. At each cross-section, there is a vertical measuring tube with a small hole connecting to the wall of the test glass tube; another vertical measuring tube with an opening located on the centerline of the glass tube facing the direction of the water flow.

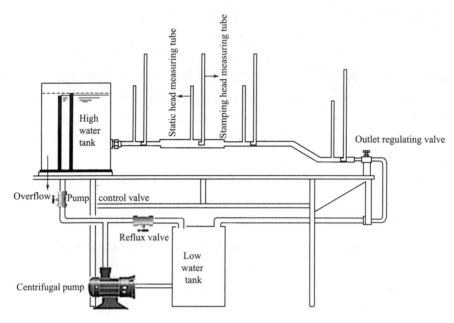

Fig. 2-4 Bernoulli equation experiment—Schematic diagram of experimental device and flow chart

图 2-4 伯努利方程实验——实验装置示意图及流程

(2) Experimental simulation interface (Fig. 2-5)

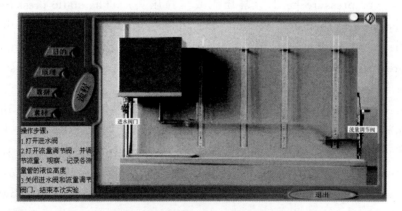

Fig. 2-5 Bernoulli experiment—Simulation interface

图 2-5 伯努利实验——仿真界面

IV. Experimental steps

(1) Experimental operations

① Fill the low tank with a certain amount of water, then close the outlet-regulating valve of the centrifugal pump, and start the centrifugal pump.

② Gradually open the outlet regulating valve of centrifugal pump, fill the upper tank until it overflows (溢流).

③ Check whether there are bubbles in the horizontal glass test tube and the vertical measuring tube. If there are bubbles, open the outlet valve of the test tube wider and let the water flow take them away. Remove bubbles in the vertical measuring tube by rubber suction ball.

④ Close the outlet-regulating valve, observe and record the liquid level height H of each test tube under the static state.

⑤ Half open the outlet regulating valve, observe and record the liquid level height H' of each section of the measuring tube.

⑥ Fully open the outlet regulating valve, observe and record the liquid level height H'' of each section of the measuring tube.

⑦ End the experiment and turn off the centrifugal pump.

(2) Operation notes

① The centrifugal pump outlet valve should not be too much open to avoid the water rushing out of the high tank and causing unstable liquid level in the high tank.

② When the outlet control valve of the test tube is widely opened, check whether the liquid level in the high tank is stable. When the liquid level drops, properly open the outlet control valve of pump.

③ Slowly turned down the outlet-regulating valve of the test tube to avoid sudden drop of flow or overflow of water in the measuring tube.

④ Check whether there are air bubbles in both the glass test tube and measuring tube, and remove the bubbles in time.

⑤ Centrifugal pump is prohibited to work in idling or with pump outlet valves fully closed (离心泵不要在空转和泵出口阀门全关的情况下工作).

V. Questions

① When the outlet valve of the conduit is fully closed, is there any change in the liquid level height in each measuring tube?

② What is the physical meaning of the liquid level height H' of measuring tube with the pressure hole facing the direction of flow?

③ Why is the farther away from the tank, the greater the ($H-H'$) difference? What is the physical meaning of this difference?

Experiment 3   Comprehensive experiment of fluid mechanics
实验 3   流体力学综合实验

The comprehensive experiment of fluid mechanics includes measuring the straight pipe resistance and local resistance of fluid flow in pipeline, verifying the flow coefficient of the flow meter, and determining the characteristic curve of centrifugal pump at certain rotate speed. All above three experiments are based on Bernoulli equation（流体力学综合实验包括流体在管路内流动时的直管和局部阻力的测定，流量计的流量系数校核和在一定的转速下离心泵的特性曲线的测定。这三个实验都以伯努利方程为基础）.

In order to overcome the resistance created by fluid flow, some mechanical energy converts into heat, which cause the temperature of the fluid to rise slightly, and cannot be used directly for fluid transport. Practically speaking, this part of the mechanical energy is lost. Σh_f in Bernoulli equation, namely represents various energy lost per kilogram of fluid to overcome various resistance to fluid flow（伯努利方程中，Σh_f 项代表每千克流体因克服各种流体流动阻力而损耗的能量）. In the practical application of Bernoulli equation, Σh_f must be calculated whether to express the conversion relationship between various energies or to calculate the energy and power required by the fluid conveyor. For long distance fluid transportation, the work done by fluid conveying machinery is mainly to overcome the fluid resistance in the pipeline. Therefore, the magnitude of resistance is related to the power consumption of fluid conveying machinery, but also related to the choice of fluid conveying machinery. The fluid resistance relates to the properties of the fluid (such as the viscosity), type of fluid flow (laminar or turbulent flow), wall conditions (rough or smooth) of the pipeline, the flow distance, and the size of the cross-section.

Flowmeter manufactured according to the standard specification, such as the orifice plate flowmeter（孔板流量计）or Venturi flowmeter（文丘里流量计）, is installed on the pipeline to measure the flow rate of fluid. A flow coefficient or a calibration curve calculated by the specified formula is generally given for operator to use the flowmeter. Some special cases often occur, for example, the original flow curve is accidentally lost, or the flowmeter is worn out after long-term use, or the composition or state of the measured flow is different from that of the standard fluid, or some non-standard forms of flowmeters are needed for scientific research. It is necessary to calibrate the flow meter and calculate the specific formula or flow curve for accurately measuring the flow.

Pump is a machinery for conveying liquid. The flow Q, head H and power N shown on the nameplate of the centrifugal pump are the values corresponding to the highest efficiency of the centrifugal pump at a certain rotate speed（离心泵铭牌上所示的流量、扬程和功率是离心泵在一定转速下效率最高点所对应的 Q，H 和 N 的值）. The head H, shaft power N and efficiency η of centrifugal pump change with the flow rate at a certain rotate speed, and the variation relationship can be expressed by the characteristic curve of the centrifugal pump（离心泵特性曲线）. Usually, the conveying capacity of centrifugal pump under a given pipeline condition can be determined according to the H-Q curve; the motor power can be reasonably selected for centrifugal pump according to the N-Q curve; and the working condition of centrifugal pump can be selected to achieve the maximum efficiency according to the η-Q curve. At present, the characteristic curve of centrifugal pump cannot be accurately calculated by analytical method, but only by experimental measurement.

Experiment 3.1 Determination of the pipeline fluid resistance
实验 3.1 管道流体阻力测定

I. Experiment purpose

① Master the method of measuring fluid resistance.

② Determine the relationship between friction coefficient of straight pipe and Reynolds number in logarithmic coordinate system.

③ Determine the equivalent length（当量长度）of the gate valve.

II. Experimental principle

When the fluid flows in the pipeline, it is inevitable to consume a certain amount of mechanical energy due to the existence of viscous shear stress and eddy current. The resistance of fluid flow in a pipe can be divided into two parts, one is the resistance due to anelasticity（内摩擦力）of fluid flowing through a straight pipe with a certain diameter, which is called the straight pipe resistance（直管阻力）. Another important part of the energy loss is the local resistance（局部阻力）. When the fluid flows through the inlet, outlet, elbow, valve, expansion, contraction and other local positions of the pipeline, both the magnitude and direction of the velocity are changed. Moreover, the fluid is disturbed, which aggravates the eddy current phenomenon and consumes energy.

（1）Resistance loss of fluid flowing in a straight pipe（流体在直管内流动时所产生的阻力损失）

The Bernoulli equation between two sections:

$$gz_1+\frac{u_1^2}{2}+\frac{p_1}{\rho}=gz_2+\frac{u_2^2}{2}+\frac{p_2}{\rho}+h_f \qquad (2\text{-}3)$$

When the fluid flows steadily in a horizontal pipe of uniform diameter（流体在直径均匀的水平管道中作稳态流动时）, the resistance loss is manifested in the reduction of pressure, that is:

$$h_f=\frac{\Delta p}{\rho}=\frac{p_1-p_2}{\rho} \qquad (2\text{-}4)$$

In other words, the resistance loss h_f of the fluid can be obtained if the pressure difference Δp between the two sections of the fluid can be measured.

The pressure difference Δp of fluid can be measured by a U-tube manometer（流体的压强差 Δp，可用 U 形管压差计测得）. The formula of U-tube manometer is:

$$\Delta p=R(\rho_0-\rho)g \qquad (2\text{-}5)$$

Where, R—U-tube manometer reading, m;

ρ_0—indicating liquid density in the pressure gauge, kg/m^3;

ρ—density of the fluid, kg/m^3;

g—gravity acceleration, $g = 9.81 \text{m/s}^2$.

The factors affecting the resistance loss are complex. In order to simplify the experiment, the variables can be combined into a number relationship by dimensional analysis method. According to dimensional analysis, the factors affecting the resistance loss are as follows: fluid properties—density ρ, viscosity μ; geometric dimensions of the fluid—pipe diameter d, pipe length L, and pipe wall roughness ε; flow condition—flow rate u.

That is:

$$\Delta p = f(d, L, u, \rho, \mu, \varepsilon)$$

The resultant dimensionless formula is as follows:

$$\frac{\Delta p}{\rho u^2} = \varphi\left(\frac{du\rho}{\mu}, \frac{L}{d}, \frac{\varepsilon}{d}\right) \tag{2-6}$$

$$\frac{\Delta p}{\rho} = \frac{L}{d}\varphi\left(Re, \frac{\varepsilon}{d}\right)\frac{u^2}{2} \tag{2-7}$$

Substituting Eq. (2-7) into Fanning equation (2-8):

$$h_f = \frac{\Delta p}{\rho} = \lambda \frac{L}{d}\frac{u^2}{2} \tag{2-8}$$

Gives:

$$\lambda = \varphi\left(Re, \frac{\varepsilon}{d}\right) \tag{2-9}$$

Where, λ is the friction coefficient (摩擦系数) of the straight pipe.

Eq. (2-8) is useful for calculating λ from a known pipe size d and velocity u converted from the measured flow rate. For laminar flow, $\lambda = 64/Re$, while the relation between λ and Re in turbulent flow is affected by the roughness of the tube wall, which needs to be obtained experimentally (层流时 $\lambda = 64/Re$；湍流时 λ 与 Re 的关系受管壁粗糙度的影响，需由实验求得).

(2) Local resistance expressed in two ways

① Local equivalent length method (局部当量长度法) The loss of energy due to local resistance to flow through a fitting or valve is equivalent to the energy loss through a straight pipe of the certain length with the same pipe diameter (流体通过某一管件或阀门时因局部阻力而造成的能量损失，相当于流体通过与其具有相同管径的若干米长度的直管能量损失). This straight pipe length is the equivalent length, denoted by L_e. In this way, the energy loss of local resistance can be calculated by the straight pipe resistance formula, namely:

$$h'_f = \lambda \frac{L_e}{d}\frac{u^2}{2} \tag{2-10}$$

From the above equation:

$$L_e = \frac{2dh_f}{\lambda u^2} \tag{2-11}$$

② Resistance coefficient method (阻力系数法) The resistance coefficient method approximately considers that the local resistance loss follows the velocity square law, namely:

$$h_f = \zeta \frac{u^2}{2} \tag{2-12}$$

Where, ζ = local resistance coefficient (dimensionless), measured by experiment.

III. Experimental device and process

(1) Schematic diagram of experimental equipment and flow chart (Fig. 2-6)

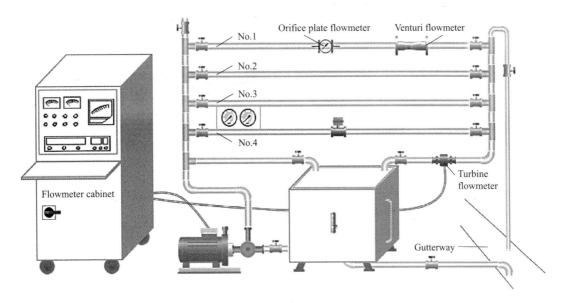

Fig. 2-6 Determination of the pipeline fluid resistance—Device schematic diagram and flow chart
图 2-6 管道流体阻力测定实验——实验装置示意图及流程

Water as the working fluid is supplied by pump circulation through a header tank (or experimental self-provided tank). The relationship between friction coefficient λ and Reynolds number Re of smooth straight pipe was measured with No. 2 pipeline [1″-galvanized iron tube, 1″-白铁管 (1″=2.54cm)], and that of a rough straight pipe was measured using No. 3 pipeline (1″-stainless steel tube, 1″-不锈钢管). The equivalent length L_e of the gate valve (full opening) is measured when the fluid flows through the No. 4 pipeline ($1\frac{1}{2}$″-galvanized iron tube, $1\frac{1}{2}$″-白铁管). The flow rate is displayed with a digital flow integrator, and the resistance loss as the fluid flows through a straight pipe or gate valve is measured using a U-tube manometer.

(2) Experimental simulation interface (Fig. 2-7)

IV. Experimental steps

(1) Experimental operations

① Familiar with the pipeline system, select the water supply mode, open the valve on

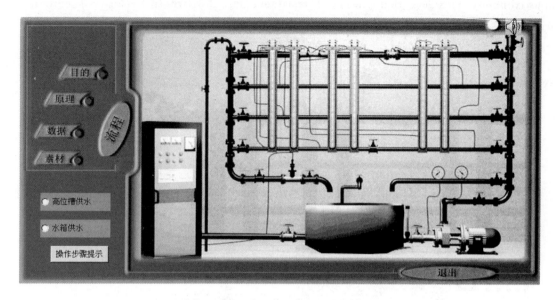

Fig. 2-7　Determination of the pipeline fluid resistance—Experimental simulation interface
图 2-7　管道流体阻力测定实验——仿真界面

the tested pipeline according to the experimental requirements and simultaneously close the valve unrelated to this experiment.

② Remove the accumulated air in the pipeline system to balance the liquid columns at both ends of U-tube manometer.

③ Determine the test items at each measuring point according to the experimental procedure.

④ Turn on the digital flow integrator, adjust the flow rate, and acquire the data after stabilization. When measuring straight pipe resistance, eight experimental data are taken reasonably between the maximum and minimum flow rate（在最大流量和最小流量之间合理地取 8 个读数）.

⑤ After experiment, close the inlet valve of the header tank or pump, and restore the valves on the comprehensive test platform.

(2) Operating notes

① When adjusting the flow rate, ensure that the fluid flow in the tube is turbulent even at the minimum flow rate（调节流量时，必须注意最小流量也要保证流体在管内流动呈湍流）.

② Acquire the data on each measuring point after stabilization.

③ Ensure that there is overflow in the header tank to keep the stability of the liquid level in the header tank.

V. Record and organize experimental data

(1) Experimental data recording（Fig. 2-8，Fig. 2-9）

① Straight pipe resistance determination（直管阻力的测定）

Chapter 2 Experiment of Chemical Engineering Principle

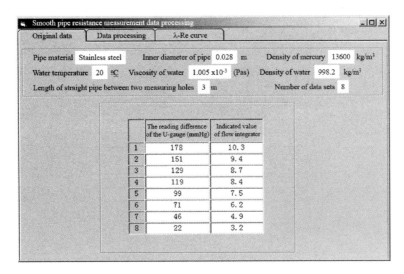

Fig. 2-8　Determination of the pipeline fluid resistance—Data recording of straight pipe resistance（1mmHg＝133.322Pa）

图 2-8　管道流体阻力测定实验——直管阻力数据记录

② Determination of equivalent length L_e of No.1 gate valve（full opening）[1″闸阀（全开时）的当量长度 L_e 的测定]

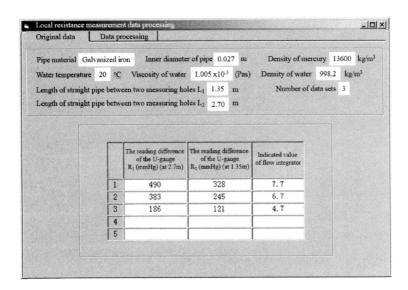

Fig. 2-9　Determination of the pipeline fluid resistance—
Data recording of valve resistance

图 2-9　管道流体阻力测定实验——闸阀阻力数据记录

（2）Experimental data collation （Fig. 2-10，Fig. 2-11）

① Straight pipe resistance determination

② Determination of equivalent length L_e of No.1 gate valve（full opening）

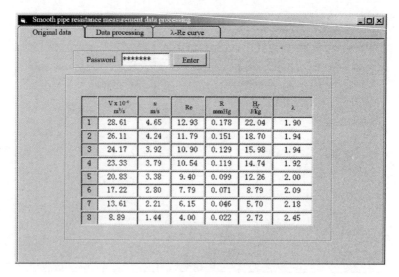

Fig. 2-10　Determination of the pipeline fluid resistance—Data collation of straight pipe resistance

图 2-10　管道流体阻力测定实验——直管阻力数据整理

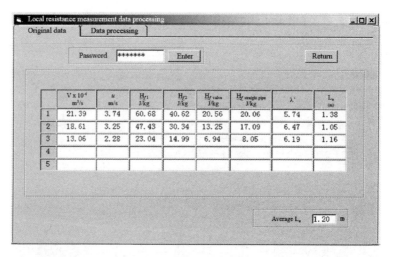

Fig. 2-11　Determination of the pipeline fluid resistance—Data collation of equivalent length L_e of No. 1 gate valve (full opening)

图 2-11　管道流体阻力测定实验——闸阀阻力数据整理

VI. Requirements for experimental report

(1) Straight pipe resistance measurement: draw λ-Re curve on log-log coordinate paper (Fig. 2-12)(直管阻力测定：在双对数坐标纸上绘制 λ-Re 曲线).

(2) Equivalent length L_e measurement of gate valves (full opening)[闸阀（全开）的当量长度 L_e 测定].

Chapter 2　Experiment of Chemical Engineering Principle

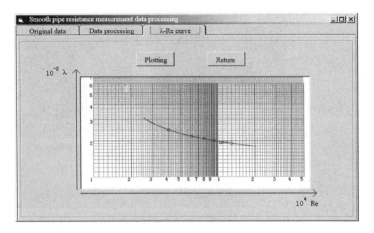

Fig. 2-12　Determination of the pipeline fluid resistance—Data plotting of straight pipe resistance
图 2-12　管道流体阻力测定实验——直管阻力数据作图

VII. Relevant materials (Fig. 2-13 to Fig. 2-17)

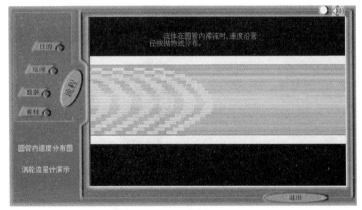

Fig. 2-13　Determination of the pipeline fluid resistance— Demonstration of velocity distribution in a circular pipe
图 2-13　管道流体阻力测定实验——圆管内速度分布演示

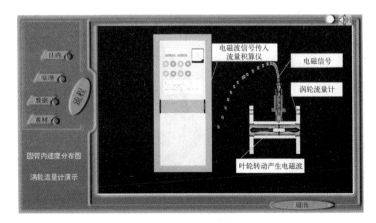

Fig. 2-14　Determination of the pipeline fluid resistance—Demonstration of turbine flowmeter
图 2-14　管道流体阻力测定实验——涡轮流量计演示

Fig. 2-15　Determination of the pipeline fluid resistance—Demonstration of gate valve
图 2-15　管道流体阻力测定实验——闸阀演示

Fig. 2-16　Determination of the pipeline fluid resistance—Demonstration of valve core
图 2-16　管道流体阻力测定实验——阀芯演示

Fig. 2-17　Determination of the pipeline fluid resistance—Demonstration of valve shell
图 2-17　管道流体阻力测定实验——阀壳演示

VIII. Questions

① What is the purpose of emptying air in the pipe before the experiment?

② How to adjust the flow rate to evenly distribute the test points on the log-log paper?

③ How to avoid measurement error when measuring the resistance loss of gate valve?

Experiment 3.2 Calibration of orifice meter and Venturi meter
实验 3.2 孔板流量计及文丘里流量计校验

I. Experiment purpose

① Familiar with the structure and application of orifice meter and Venturi meter.

② By measuring the flow coefficient of orifice and Venturi meter, master the calibration method of flowmeter.

II. Experimental principle

When the fluid steadily flows through the orifice (orifice plate or Venturi tube throat) in the pipeline, the reduction of the cross-section area increases the velocity head at the expense of the pressure head. The higher the flow rate, the greater the pressure drop. The flow coefficient of orifice flowmeter or Venturi flowmeter can be measured by this principle〔在管路中作稳定流动的流体通过孔口（孔板或文丘里管喉道）时，由于截面积缩小，流速增大，静压能降低而造成孔口前后有一定的压降，流速愈大，压降愈大，由此原理可测得孔板流量计或文丘里流量计的流量系数〕。

(1) Orifice meter（孔板流量计）

An orifice meter consists of an accurately machined and drilled plate mounted between two flanges with the hole concentric with the pipe in which it is mounted. The pressure taps before and after the orifice connect with the two arms of the U-tube manometer. When the fluid flows through the orifice of the orifice plate, the pressure drop generates due to the velocity change. The contraction occurs at the outlet to form the "vena contracta（流颈）", where the flow channel section is the smallest and the flow velocity is the largest, and the static pressure drop is corresponding the largest. Orifice meter measures fluid flow according to pressure drop as a function of flow.

The basic equation for an orifice meter is obtained by deducing Bernoulli equation of incompressible fluids in the pipeline. Temporarily omitting the energy loss, the Bernoulli equation at the upstream section of the orifice plate and the vena contracta is:

$$\frac{u_2^2 - u_1^2}{2} = \frac{p_1 - p_2}{\rho} \tag{2-13}$$

or

$$\sqrt{u_2^2 - u_1^2} = \sqrt{\frac{2(p_1 - p_2)}{\rho}} \tag{2-14}$$

Since the location and section area of the vena contracta are difficult to determine, while the aperture of the orifice plate is known, the velocity u_2 of the vena contracta can be replaced with the orifice velocity u_0 of the orifice plate（由于流颈处的面积很难确定，而孔

板的孔径是已知的，用孔板的孔口流速 u_0 代替流颈处的流速 u_2）. On the other hand, in the derivation of Eq. (2-14), the energy loss between the two sections is temporarily omitted. To account for the error caused by neglecting the energy loss, Eq. (2-14) is corrected by introducing a correction coefficient C:

$$\sqrt{u_0^2 - u_1^2} = C\sqrt{\frac{2(p_1 - p_2)}{\rho}} \tag{2-15}$$

For incompressible fluids, according to the continuity equation

$$u_1 = u_0 A_0 / A_1 \tag{2-16}$$

Substituting Eq. (2-16) into Eq. (2-15) to get

$$u_0 = \frac{C\sqrt{2(p_1 - p_2)/\rho}}{\sqrt{1 - \left(\frac{A_0}{A_1}\right)^2}} \tag{2-17}$$

Let

$$C_0 = \frac{C}{\sqrt{1 - \left(\frac{A_0}{A_1}\right)^2}} \tag{2-18}$$

The pressure drop at front and rear position of the orifice plate is measured with a U-tube manometer, namely:

$$p_1 - p_2 = R(\rho_0 - \rho)g \tag{2-19}$$

Then

$$u_0 = C_0 \sqrt{\frac{2R(\rho_0 - \rho)g}{\rho}} \tag{2-20}$$

The volumetric flow rate V_s can be obtained from the velocity u_0 (m/s) and the cross-sectional area A_0 of the orifice:

$$V_s = u_0 A_0 = C_0 A_0 \sqrt{\frac{2gR(\rho_0 - \rho)}{\rho}} \tag{2-21}$$

Where, A_0—area of orifice, m^2;

g—acceleration of gravity, m/s^2;

C_0—coefficient, dimensionless;

R—U-tube manometer reading, m;

ρ—density of fluid, kg/m^3;

ρ_0—density of indicating liquid of U-tube manometer, kg/m^3.

The flow coefficient of the orifice plate relates to the energy loss of the fluid flowing through the orifice plate, the position of the pressure tap, the ratio of aperture to pipe diameter and Reynolds number. When the A_0/A_1 is fixed and Reynolds number is greater than a certain value, C_0 is approximately a constant value（当 A_0/A_1 一定时，雷诺数 Re 超过某个数值后，C_0 就接近于定值）.

(2) Venturi meter（文丘里流量计）

The orifice meter has certain practical disadvantages for the serious energy loss caused by the eddy current when the fluid flows through the orifice. A Venturi meter has a short conical inlet section leading to a throat section, then to a long discharge cone. Such structure

Chapter 2 Experiment of Chemical Engineering Principle

can largely avoid the loss caused by eddy current（文丘里流量计为一管径渐渐均匀缩小而后又渐渐均匀扩大的光滑管子，这样可以在很大程度上避免涡流所产生的损失）. The principle of the Venturi meter is identical with that of the orifice meter.

The flow rate at the pipe throat is:

$$V_s = u_0 \times A_0 = C_v A_0 \sqrt{\frac{2gR(\rho_0-\rho)}{\rho}} \tag{2-22}$$

In the equation, C_v is the flow coefficient of Venturi meter, and other items are of the same physical meanings as those of orifice meter.

III. Experimental device and process

(1) Schematic diagram of experimental equipment and flow chart

Shown as Fig. 2-6.

Water as the working fluid is supplied by pump circulation through a header tank (or experimental self-provided tank). The fluid flows through the orifice meter and Venturi meter in No. 1 pipeline successively, and the coefficient C_0 and C_v of both meters are calibrated experimentally. The static pressure drop generated by the flow through the orifice plate and the Venturi meter is measured with a U-tube manometer, and the flow rate is displayed by a digital flow integrator.

(2) Experimental simulation interface

Shown as Fig. 2-7.

IV. Experimental steps and operating notes

Refer to fluid resistance measurement experiment.

V. Record and collation of experimental data (Fig. 2-18, Fig. 2-19)

Experimental data recording and processing

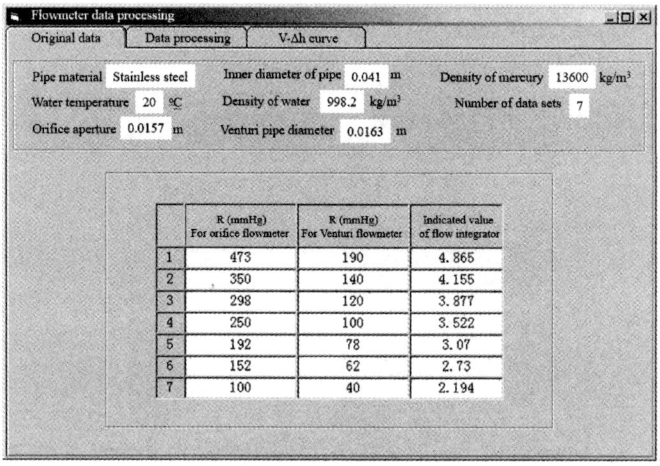

Flowmeter data processing

Original data | Data processing | V-Δh curve

Pipe material: Stainless steel   Inner diameter of pipe: 0.041 m   Density of mercury: 13600 kg/m³
Water temperature: 20 ℃   Density of water: 998.2 kg/m³   Number of data sets: 7
Orifice aperture: 0.0157 m   Venturi pipe diameter: 0.0163 m

	R (mmHg) For orifice flowmeter	R (mmHg) For Venturi flowmeter	Indicated value of flow integrator
1	473	190	4.865
2	350	140	4.155
3	298	120	3.877
4	250	100	3.522
5	192	78	3.07
6	152	62	2.73
7	100	40	2.194

Fig. 2-18 Calibration of orifice meter and Venturi meter—Experimental data recording

图 2-18 孔板流量计及文丘里流量计校验实验——数据记录

	V×10⁻⁴ m³/s	Orifice flowmeter				Venturi flowmeter			
		u_1 m/s	R m/s	Δh J/kg	C_0	u_2 m/s	R m/s	Δh J/kg	C_v
1	13.51	6.98	0.473	58.58	0.64	6.48	0.190	23.53	0.94
2	11.54	5.96	0.350	43.35	0.64	5.53	0.140	17.34	0.94
3	10.77	5.56	0.298	36.91	0.65	5.16	0.120	14.86	0.95
4	9.78	5.05	0.250	30.96	0.64	4.69	0.100	12.38	0.94
5	8.53	4.41	0.192	23.78	0.64	4.09	0.078	9.66	0.93
6	7.58	3.92	0.152	18.82	0.64	3.63	0.062	7.68	0.93
7	6.09	3.15	0.100	12.38	0.63	2.92	0.040	4.95	0.93

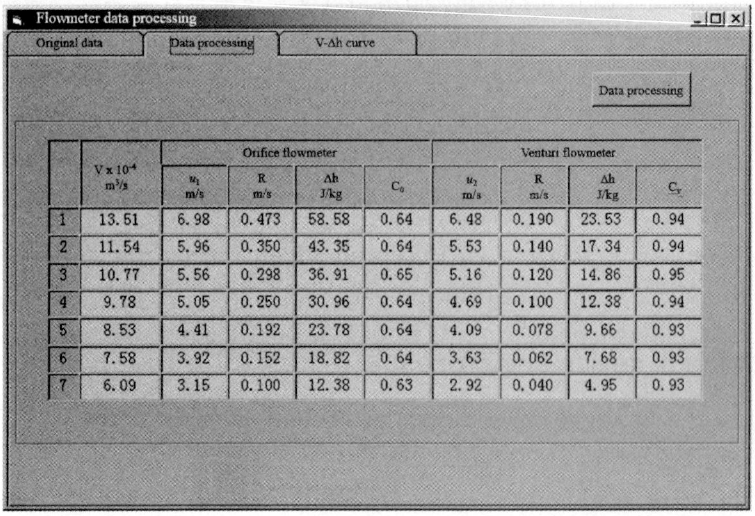

Fig. 2-19 Calibration of orifice meter and Venturi meter—Experimental data processing

图 2-19 孔板流量计及文丘里流量计校验实验——数据整理

Ⅵ. Requirements for experimental report

(1) On the log-log coordinate paper, plot straight lines of flow rate V_s versus static pressure drop Δh $\left[\Delta h=\dfrac{gR(\rho_0-\rho)}{\rho}\right.$, (J/kg) $\left.\right]$ with slope of 1/2 for orifice meter and Venturi meter, respectively（在双对数坐标纸上分别作出孔板流量计与文丘里流量计斜率为 1/2 的 V-Δh 直线）(Fig. 2-20). Then calculate the flow coefficient C_0 of the orifice meter and the flow coefficient C_v of the Venturi meter.

(2) Calculate the flow coefficient C_0 of the orifice meter and the flow coefficient C_v of the Venturi meter by mean value method.

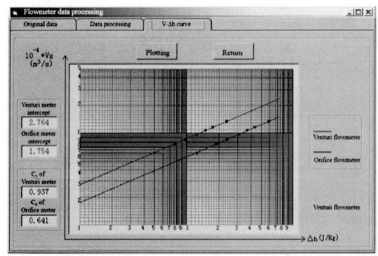

Fig. 2-20 Calibration of orifice meter and Venturi meter—Experimental data plotting

图 2-20 孔板流量计及文丘里流量计校验实验——数据作图

VII. Relevant materials (Fig. 2-21, Fig. 2-22)

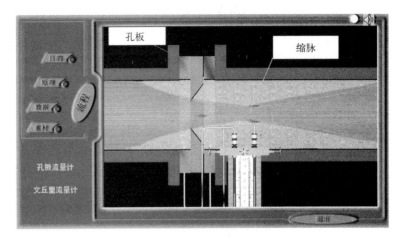

Fig. 2-21 Calibration of orifice meter and Venturi meter—Demonstration of orifice meter
图 2-21 孔板流量计及文丘里流量计校验实验——孔板流量计演示

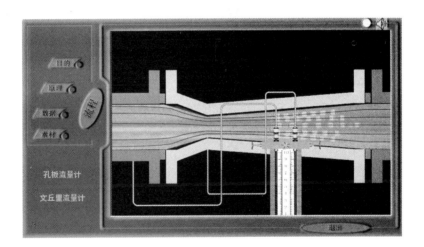

Fig. 2-22 Calibration of orifice meter and Venturi meter—Demonstration of Venturi meter
图 2-22 孔板流量计及文丘里流量计的校验实验——文丘里流量计演示

VIII. Questions

① At the same flow rate, compare the pressure difference shown by orifice meter and Venturi meter respectively, analyze the reasons.

② To get the flow coefficient, what items were calibrated in the experiment?

③ Why is the V-Δh calibration graph of orifice meter or Venturi meter a straight line with slope of 1/2 on a log-log coordinate paper?

Experiment 3.3 Determination of the centrifugal pump performance curve
实验3.3 离心泵性能曲线的测定

I. Experiment purpose

① Master the structure, performance and the operation of centrifugal pump.

② Plot the performance curve of centrifugal pump at certain rotate speed.

II. Experimental principle

See Experiment 4.

III. Experimental device and process

(1) Schematic diagram of experimental equipment and flow chart

See Fig. 2-6.

After merging the water in the self-provided water tank by the pump through the No. 1 pipeline with any one from the No. 2, No. 3 and No. 4 pipelines, the water is returned to the own water tank through the turbine flow transmitter and control valve to form the water circulation system. The flow rate is displayed with a digital flow accumulator. The power, vacuum and pressure are measured by a power meter, vacuum meter and pressure gauge respectively.

(2) Experimental simulation interface

See Fig. 2-7.

IV. Experimental steps

(1) Experimental operations

① Close the release valve of the self-provided water tank of the test bench and drain water to about 2/3 of the water tank.

② Open the valve on the No. 1 pipe and any one of the No. 2, No. 3 and No. 4 pipes. Close the pressure-measuring valve of the U-tube manometer on the pipe to prevent the mercury from flushing out.

③ Check the motor and the centrifugal pump for normal operation. Turn on the power switch of the motor (observe the operation of the motor and the centrifugal pump, and cut off the power supply immediately in case of abnormality).

④ During the experiment, gradually open the flow control valve to increase the flow, and measure 6-8 sets of data.

⑤ At the end of the experiment, stop the pump and restore the pipeline valve.

(2) Operating notes

① Before starting the pump, close the pipe control valve. In particular, close the pressure-measuring valve of the U-tube manometer on the pipeline through which the fluid flows.

② Reasonably divide the flow points within the maximum flow range, and adjust the flow by the control valve.

③ After adjusting the flow, read the data of each parameter after stabilizing. And the reading of zero flow cannot be omitted.

V. Record and collation of experimental data

(1) Experimental data recording

See Experiment 4.

(2) Experimental data collation

See Experiment 4.

VI. Requirements for experimental report

Draw the characteristic curve of centrifugal pump on ordinary coordinate paper.

VII. Relevant materials

See Experiment 4.

VIII. Questions

See Experiment 4.

Experiment 4　Determination of the characteristic curves of a centrifugal pump
实验4　离心泵的特性曲线的测定

I. Experiment purpose

① Master the structure, performance and the operation of centrifugal pump.

② Plot the performance curve of centrifugal pump at certain rotational speed.

③ Plot the pipeline characteristic curve.

④ Understand the operating point and flow regulation of centrifugal pump.

II. Experimental principle

(1) Characteristic curve of centrifugal pump

The performance parameters of centrifugal pump depend on the internal structure of pump, impeller form and rotational speed（离心泵的性能参数取决于泵的内部结构、叶轮形式及转速）.

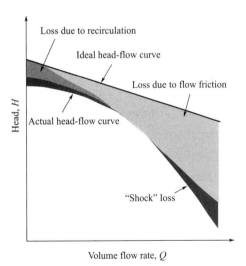

Fig. 2-23　Theoretical and actual characteristic curve of a centrifugal pump

图 2-23　离心泵的理论压头与实际压头

The relationship between theoretical head and flow rate can be obtained by theoretical analysis of liquid particle motion in pump, as shown in Fig. 2-23. The developed head of an actual pump is less than that calculated from the ideal pump relation primarily resulted from circulatory flow and fluid friction, etc. Many complicating factors determine the actual efficiency and performance characteristics of a pump. Hence, the actual experimental performance of the pump is usually employed. The plots of actual head, total power consumption, and efficiency vs. volumetric flow rate are called the characteristic curves of a pump（离心泵的特性曲线）.

① The head of pump, H_e（泵的扬程）

Set up Bernoulli equation between the vacuum gauge and pressure gauge located at the inlet and the outlet pipes of the centrifugal pump (unit weight of liquid as the calculation standard), then

$$z_1 + \frac{p_1}{\rho g} + \frac{u_1^2}{2g} + H_e = z_2 + \frac{p_2}{\rho g} + \frac{u_2^2}{2g} + H_f \tag{2-23}$$

Because the two pressure taps（测压孔）are close to the inlet and outlet of the centrifugal pump, the friction loss of the straight section is very small, the resistance loss is classified into the efficiency of the centrifugal pump, so $H_f = 0$.

$$H_e = \frac{p_2}{\rho g} - \frac{p_1}{\rho g} + h_0 + \frac{u_2^2 - u_1^2}{2g} \tag{2-24}$$

With the same inlet and outlet diameter of the centrifugal pump, $u_1 = u_2$.

Substituting it into Eq. (2-24) gives:

$$H_e = \frac{p_2}{\rho g} - \frac{p_1}{\rho g} + h_0 \tag{2-25}$$

$$H_e = H_{\text{pressure gauge}} + H_{\text{vacuum gauge}} + h_0 \tag{2-26}$$

Where, $H_{\text{pressure gauge}}$—gauge pressure expressed in liquid column height, m;

$H_{\text{vacuum gauge}}$—vacuum expressed in liquid column height, m;

h_0—0.1m, the vertical distance between the center of the pressure and the vacuum gauge.

② Shaft power of pump N_{shaft}

The actual power obtained by the centrifugal pump from the motor (that is, the work input by the motor to the centrifugal pump per unit time) is called the shaft power of the centrifugal pump（离心泵的轴功率）.

The relationship between the shaft power of the pump and the electric power of the motor is as follows:

$$N_{\text{shaft}} = N_{\text{motor}} \eta_{\text{motor}} \eta_{\text{transmission}} \tag{2-27}$$

Where, N_{shaft}—electric power of the motor, measured by the power meter, kW;

η_{motor}—motor efficiency, η_{motor} is 0.9 in the experiment;

$\eta_{\text{transmission}}$—transmission efficiency, $\eta_{\text{transmission}}$ is 1.0 in the experiment.

③ Efficiency of pump η

The efficiency of pump is the ratio of the effective power delivered to the fluid and shaft power of centrifugal pump. It is denoted by η and defined by

$$\eta = \frac{N_e}{N_{shaft}} \times 100\% \tag{2-28}$$

$$N_e = \frac{H_e Q \rho}{102} \tag{2-29}$$

Where, N_e—effective power of centrifugal pump, kW;

Q—flow rate, m³/s;

H_e—pump head, m;

ρ—density of fluid, kg/m³.

(2) Pipeline characteristic curve of the centrifugal pump（管路特性曲线）

When the centrifugal pump is installed in a specific pipeline system, its actual working pressure head and flow are not only related to the performance of the centrifugal pump itself, but also related to the pipeline characteristics. In other words, the pump and the pipeline are mutually restricted during fluid delivery.

For a particular pipe system, it can be derived from the Bernoulli equation:

$$H_e = K + BQ^2 \tag{2-30}$$

Where, H_e—pressure head required by pipeline, m;

Q—flow rate, m³/s.

Under certain operating conditions, both K and B are constant:

$$K = \Delta Z + \frac{\Delta p}{\rho g} \tag{2-31}$$

$$B = \left(\lambda \frac{L + \Sigma L_e}{d} + \Sigma \zeta \right) \frac{1}{2g(3600A)^2} \tag{2-32}$$

Where, A—cross-sectional area of the pipeline, m²;

d—pipe diameter, m;

L—pipe length, m;

L_e—local resistance equivalent length, m;

ζ—coefficient of local resistance;

ΔZ—potential energy difference, J/kg;

Δp—static pressure difference, Pa.

From the above formula, when conveying liquid in a specific pipeline, the pressure head H_e required by the pipeline changes with the square of liquid flow Q (turbulent state). If this relation is plotted on the corresponding coordinate paper, the resulting H_e-Q curve is called the pipeline characteristic curve（管路特性曲线）. The shape of the curve is dependent on the coefficients K and B, that is, on the operating conditions and the geometry of the line, but independent of the performance of the pump.

Due to the difficulty in determining K and B, the laboratory does not use this method to get the pipeline characteristic curve.

(3) Determination of pipeline characteristic curve and adjustment of working point

The centrifugal pump works in a certain pipeline system, so the head and flow rate provided by the pump must be consistent with the pressure head and flow rate required by the pipeline. If the characteristic curve H-Q of the centrifugal pump and the characteristic curve H_e-Q of the pipeline are plot on the same coordinate chart, the intersection of the two curves is called the working point (工作点) of the pump in the pipeline. If the production task changes or the selected pump does not meet the requirements of the delivery task, it is necessary to adjust the working point of the pump either by resizing the valve opening to change the pipeline characteristic curve, or by adjusting the pump speed to change the pump characteristic curve. Either method can achieve the purpose of adjusting the working point of the centrifugal pump.

When measuring the pipeline characteristic curve, if the valve opening of the centrifugal pump is fixed, the pipeline characteristic curve is certain at this time. The experimental device applies a frequency conversion governor to change the frequency of the resistance, and then change the rotational speed of the pump accordingly. Measure the flow rate at different pump speed, and record the readings of the corresponding pressure gauge, vacuum gauge and power meter. After calculating the pump head H_e (namely the pressure head required by the pipeline), draw the pipeline characteristic curve.

III. Experimental device and process

(1) Schematic diagram of experimental equipment and flow chart (Fig. 2-24)

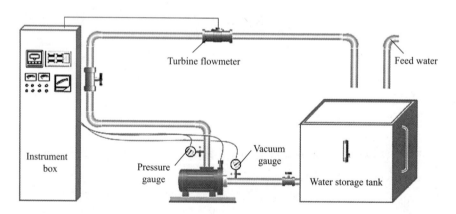

Fig. 2-24 Determination of the characteristic curves of a centrifugal pump—
Device schematic diagram and flow chart

图 2-24 离心泵特性曲线测定实验——装置流程示意图

The experimental equipment is composed of a $1\frac{1}{2}$ BA (or $1\frac{1}{2}$ BL) type centrifugal pump (Both BA and BL type are single-stage single-suction cantilever centrifugal pumps, and BL type can be regarded as the improved type of BA) under test, a water tank, pipelines, control valves, a turbine meter, a vacuum gauge and a pressure gauge, etc. The instrument cabinet is equipped with pump switch button, power meter and meter digital display instrument. See Appendix 2 for the usage of the power meter.

(2) Experimental simulation interface (Fig. 2-25)

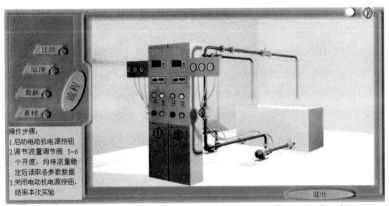

Fig. 2-25 Determination of the characteristic curves of a centrifugal pump—
Experimental simulation interface

图 2-25 离心泵特性曲线测定实验——实验仿真界面

(3) Schematic diagram of experimental equipment Ⅱ and flow chart (Fig. 2-26)

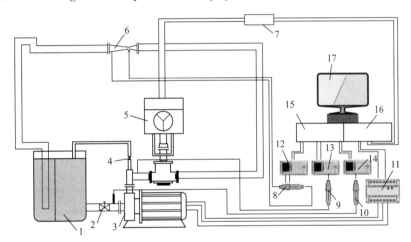

Fig. 2-26 Determination of the characteristic curves of a centrifugal pump—
Device (Ⅱ) schematic diagram and flow chart

图 2-26 离心泵特性曲线测定实验——装置Ⅱ流程示意图

1—Water storage tank (贮水箱); 2—Pump inlet control valve (泵入口调节阀); 3—Centrifugal pump (离心泵);
4—Backflow valve (回流阀); 5—Regulating valve (调节阀); 6—Venturi flow meter (文丘里流量计); 7—Relay (继电器);
8,9—Pressure sensor (压力传感器); 10—Vacuum sensor (真空度传感器); 11—Power transmitter (功率变送器);
12-14—Amplifier (放大器); 15—AD converter (AD 转换器); 16—DA converter (DA 转换器); 17—Computer (计算机)

The centrifugal pump draws the water from the storage tank into the experimental system. The flow rate is controlled with the outlet automatic control valve. The fluid flows through the conveying line to the Venturi meter to measure the flow rate, and then through the return line to the storage tank for circulation flow. The experimental device can be manually operated, and it can also realize computer data acquisition and automatic control operation.

IV. Experimental steps

(1) Experimental operations

① Check whether the motor and centrifugal pump are running normally. Turn on the power switch of the motor to observe the operation of the motor and centrifugal pump. Cut off the power immediately if there is any abnormality.

② Gradually opened the flow control valve to increase the flow rate for measuring 6-8 sets of data. Record the flow rate, motor power, gauge pressure and vacuum pressure at different flows（记录不同流量条件下的流量、电机功率、表压和真空度）.

③ Shut-off the pump when the experiment is over.

(2) Operating notes

① Close the outlet valve of the pump before starting.

② Reasonably distribute the experimental points of the flow rate within the maximum flow range（在最大流量范围内合理分割流量进行实验布点）. Adjust the flow rate by the control valve.

③ Read the data of each parameter after the adjusted flow rate is stable. In particular, it is necessary to record the readings when the flow rate is zero（不要忘记流量为零时各读数的记录）.

V. Record and organize experimental data

(1) Experimental data recording (Fig. 2-27)

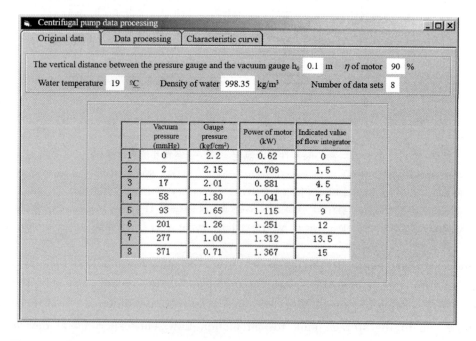

Fig. 2-27 Determination of the characteristic curves of a centrifugal pump—Data recording

图 2-27 离心泵特性曲线测定实验——数据记录

(2) Experimental data collation (Fig. 2-28)

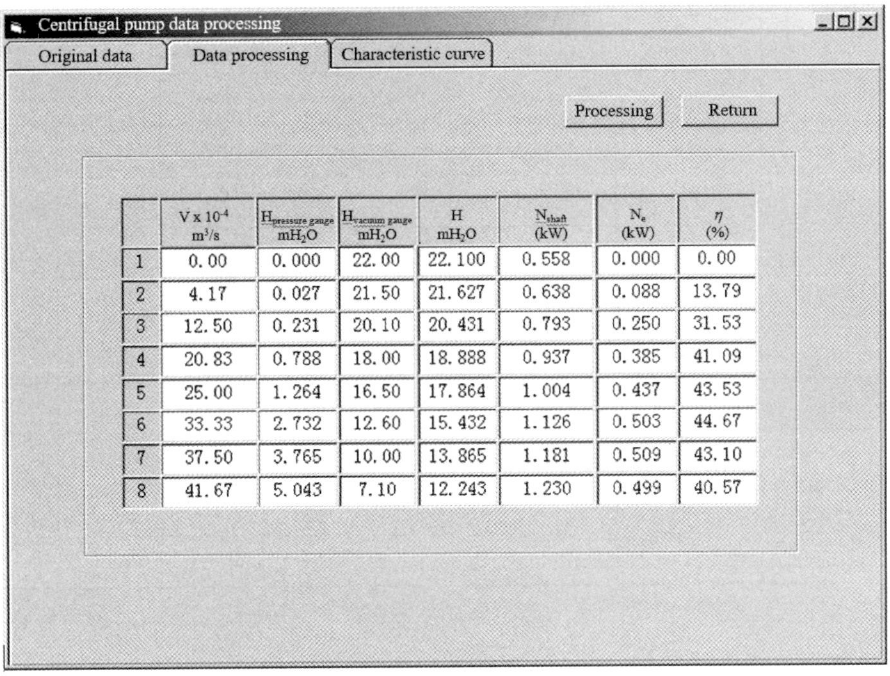

No.	$V \times 10^{-4}$ m³/s	$H_{pressure\ gauge}$ mH₂O	$H_{vacuum\ gauge}$ mH₂O	H mH₂O	N_{shaft} (kW)	N_e (kW)	η (%)
1	0.00	0.000	22.00	22.100	0.558	0.000	0.00
2	4.17	0.027	21.50	21.627	0.638	0.088	13.79
3	12.50	0.231	20.10	20.431	0.793	0.250	31.53
4	20.83	0.788	18.00	18.888	0.937	0.385	41.09
5	25.00	1.264	16.50	17.864	1.004	0.437	43.53
6	33.33	2.732	12.60	15.432	1.126	0.503	44.67
7	37.50	3.765	10.00	13.865	1.181	0.509	43.10
8	41.67	5.043	7.10	12.243	1.230	0.499	40.57

Fig. 2-28 Determination of the characteristic curves of a centrifugal pump—Data processing

图 2-28　离心泵特性曲线测定实验——数据整理

(3) Experimental data and collation

① Data of centrifugal pump characteristic curve of experimental device Ⅱ

No.	Δp of flowmeter /kPa	Pump outlet pressure/MPa	Pump inlet pressure/MPa	Power meter /kW	Head/m	Flowmeter flow/(m³/h)	Shaft power /kW	Efficiency /%
1	0	0.215	0	0.78	22.096	0.00	0.624	0.0
2	3.0	0.214	0	1.08	21.994	4.32	0.864	30.0
3	14.9	0.205	0	1.34	21.077	9.64	1.072	51.6
4	22.1	0.194	0	1.45	19.955	11.74	1.160	55.0
5	35.0	0.175	0	1.59	18.018	14.77	1.272	57.0
6	44.1	0.163	0	1.66	16.795	16.58	1.328	57.1
7	50.8	0.150	0.001	1.75	15.572	17.80	1.400	53.9
8	60.0	0.137	0.004	1.81	14.553	19.34	1.448	52.9
9	74.6	0.118	0.006	1.90	12.820	21.57	1.520	49.5
10	87.2	0.095	0.009	1.99	10.781	23.32	1.592	43.0

② Data of pipe line characteristic curve of experimental device II

No.	Motor frequency /Hz	Motor speed /(r/min)	Δp of flowmeter /kPa	Pump outlet pressure/MPa	Pump inlet pressure/MPa	Head/m	Flowmeter flow /(m³/h)
1	50	2904	55.6	0.14	0.0038	14.60	18.6
2	48	2800	52.3	0.13	0.0030	13.50	18.0
3	46	2690	48.7	0.12	0.0032	12.50	17.4
4	44	2584	44.6	0.11	0.0030	11.50	16.6
5	42	2474	40.8	0.10	0.0028	10.40	15.9
6	40	2357	36.2	0.09	0.0022	9.39	15.0
7	35	2066	28.2	0.07	0.0022	7.35	13.2
8	30	1776	20.8	0.05	0.0020	5.30	11.3
9	20	1188	8.7	0.022	0.0020	2.44	7.4
10	10	593	0	0	0.0000	0.00	0.0

VI. Requirements for experimental report

Plot the characteristic curve of centrifugal pump on the ordinary coordinate paper (Fig. 2-29 to Fig. 2-31)（在普通坐标纸上绘制离心泵的特性曲线）.

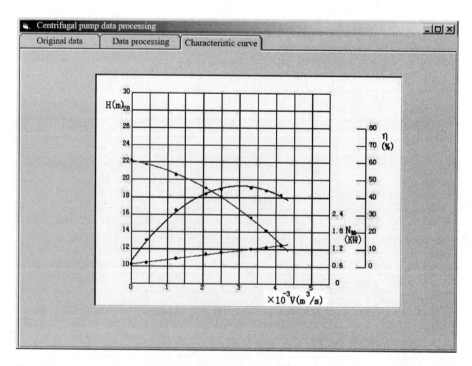

Fig. 2-29 Determination of the characteristic curves of a centrifugal pump—Data plotting

图 2-29 离心泵特性曲线测定实验——数据作图

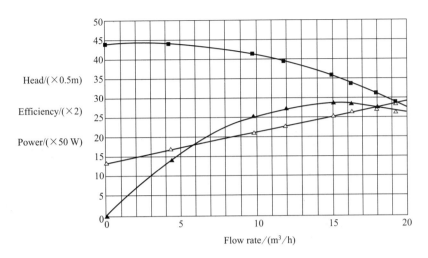

Fig. 2-30 Determination of the characteristic curves of a centrifugal pump—
Centrifugal pump characteristic curve of experimental device II

图 2-30 离心泵特性曲线测定实验——实验装置Ⅱ的离心泵特性曲线图

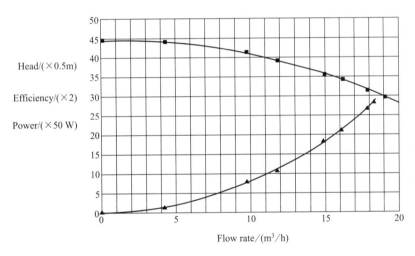

Fig. 2-31 Determination of the characteristic curves of a centrifugal pump—
Pipeline characteristic curve of experimental device II

图 2-31 离心泵特性曲线测定实验——实验装置Ⅱ的管路特性曲线图

VII. Relevant materials (Fig. 2-32 to Fig. 2-38)

VIII. Questions

① What are the possible reasons why the centrifugal pump still cannot start after pump priming?

② Why must the outlet valve be closed when the centrifugal pump starts?

③ What is the basis for choosing a centrifugal pump?

④ Try to analyze the phenomena of air-binding and cavitation.

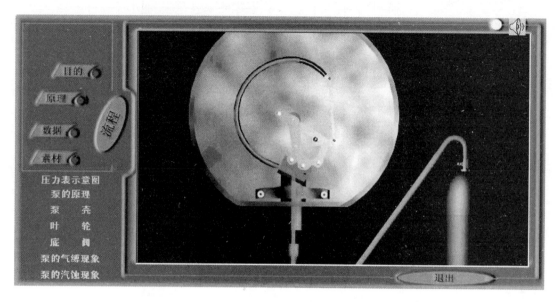

Fig. 2-32 Determination of the characteristic curves of a centrifugal pump—
Schematic diagram of pressure gauge

图 2-32 离心泵特性曲线测定实验——压力表示意图

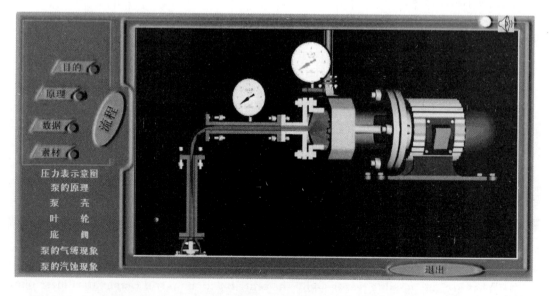

Fig. 2-33 Determination of the characteristic curves of a centrifugal pump—The principle of pump

图 2-33 离心泵特性曲线测定实验——泵的原理

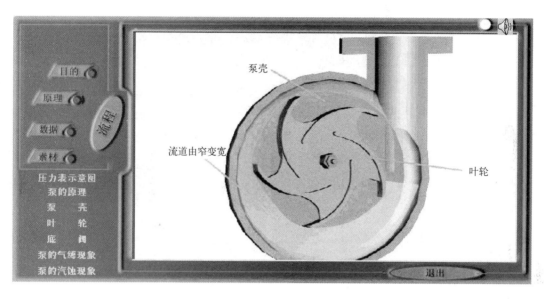

Fig. 2-34　Determination of the characteristic curves of a centrifugal pump—Pump casing
图 2-34　离心泵特性曲线测定实验——泵壳

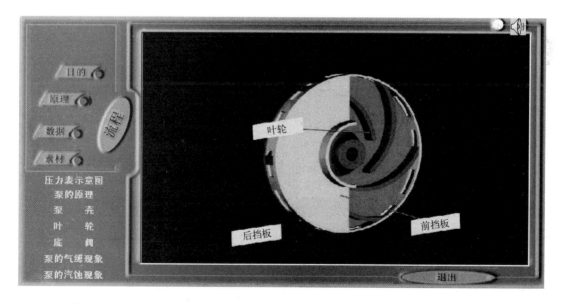

Fig. 2-35　Determination of the characteristic curves of a centrifugal pump—Impeller
图 2-35　离心泵特性曲线测定实验——叶轮

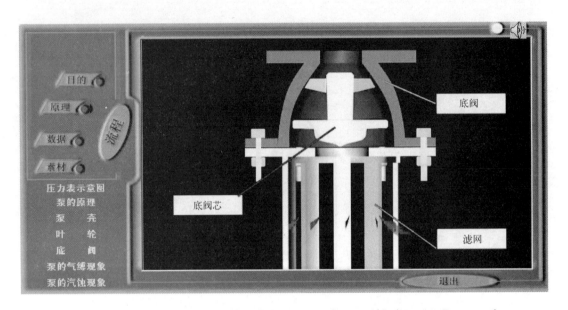

Fig. 2-36　Determination of the characteristic curves of a centrifugal pump—Bottom valve
图 2-36　离心泵特性曲线测定实验——底阀

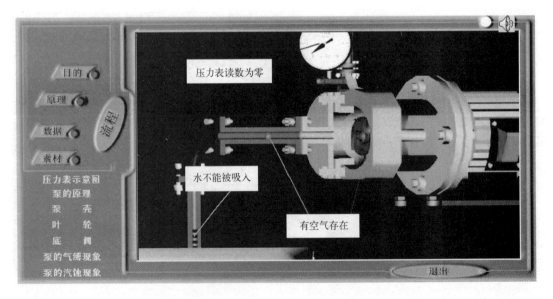

Fig. 2-37　Determination of the characteristic curves of a centrifugal pump—Air-binding of pump
图 2-37　离心泵特性曲线测定实验——泵的气缚现象

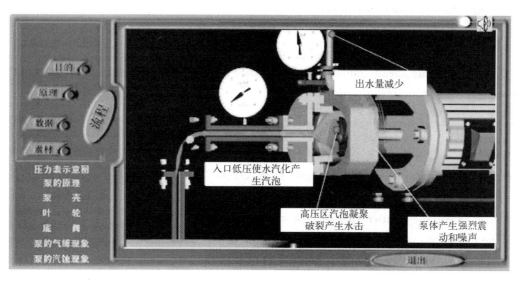

Fig. 2-38 Determination of the characteristic curves of a centrifugal pump—Pump cavitation

图 2-38 离心泵特性曲线测定实验——泵的汽蚀现象

⑤ What are the benefits and disadvantages of regulating the flow rate of centrifugal pump by changing the valve opening on the pump outlet pipe? Is there any other way to regulate the flowrate of the pump?

Experiment 5 Comprehensive experiment of convective heat transfer
实验 5 对流传热综合实验

I. Experiment purpose

① Measure the heat-transfer coefficient α_i of air-vapor system by forced convection in the ordinary and enhanced double-pipe exchangers, respectively. And induct the measured data into a dimensionless equation.

② Calculate the heat-transfer enhancement ratio by determining the ordinary and enhanced double-pipe exchangers under the same flow rate.

③ Understand the method to measure the wall temperature by thermocouple.

II. Experimental principle

(1) Determination of convective heat-transfer coefficient α_i （对流传热系数 α_i）

The convective heat transfer coefficient can be measured experimentally according to Newton's cooling law（牛顿冷却定律）：

$$\alpha_i = \frac{Q}{\Delta t_m A_i} \tag{2-33}$$

Where，α_i—convective heat transfer coefficient of air in the pipe，W/(m^2 · ℃)；

Q—heat transfer rate in pipe, W;

A_i—heat exchange area in the pipe, m²;

Δt_m—logarithmic mean temperature difference, ℃.

The logarithmic mean temperature difference（对数平均温度差）is obtained by the following formula:

$$\Delta t_m = \frac{(t_w - t_1) - (t_w - t_2)}{\ln \frac{t_w - t_1}{t_w - t_2}} \qquad (2\text{-}34)$$

Where, t_1—temperature of entering cold fluid (air), ℃;

t_2—temperature of leaving cold fluid (air), ℃;

t_w—average temperature of pipe wall, ℃.

The inner thin pipe of the double-pipe exchanger is made of red copper pipe with large thermal conductivity. Therefore, it is considered that the inner wall temperature, outer wall temperature and average surface temperature are approximately equal, which is expressed by t_w.

t_w is measured by a digital millivolt meter, and the corresponding thermoelectric potential $E(\text{mV})$ is obtained by the following formulas:

Heat transfer equipment No. 1: $t_w = 8.5 + 21.26E$ $\qquad (2\text{-}35)$

Heat transfer equipment No. 2: $t_w = 1.2705 + 23.518E$ $\qquad (2\text{-}36)$

Heat-transfer area:

$$A_i = \pi d_i L_i \qquad (2\text{-}37)$$

Where, d_i—the inner diameter of inner pipe, m;

L_i—length of inner pipe, m.

For heat-transfer equipment No. 1: $d_i = 0.01925\text{m}$; $L_i = 1.10\text{m}$; for heat-transfer equipment No. 2: $d_i = 0.02\text{m}$; $L_i = 1.00\text{m}$.

According to the energy-balance equation,

$$Q = WC_p (t_1 - t_2) \qquad (2\text{-}38)$$

Where, W—mass flow rate of air, kg/h;

C_p—specific heat capacity of air at qualitative temperature, kJ/(kg·℃);

t_1—temperature of entering cold fluid (air), ℃;

t_2—temperature of leaving cold fluid (air), ℃.

And

$$W = V\rho$$

Where, V—the average volume flow rate of air through the measurement section (that is, the volume flow rate of air under experimental conditions), m³/h;

ρ—density of air at qualitative temperature, kg/m³.

$$V = V_{t_1} \times \frac{273 + \overline{t_1}}{273 + t_1} \qquad (2\text{-}39)$$

Where, $\overline{t_1}$—the mean temperature of entering air and leaving air, $\overline{t_1} = \frac{t_1 + t_2}{2}$;

V_{t_1} — the volume flow (m³/h) under the air entering temperature (i.e., the temperature of the flowmeter composed of orifice plate, pressure sensor and digital display instrument).

Using measured Δp (kPa), V_{t_1} can be obtained by the equation

Heat transfer equipment No. 1:
$$V_{t_1} = C_0 A_0 \sqrt{\frac{2\Delta p \times 1000}{\rho_{t_1}}} \times 3600 \tag{2-40}$$

Heat transfer equipment No. 2:
$$V_{t_1} = 18.703 \times (\Delta p)^{0.563} \tag{2-41}$$

Where, C_0 — coefficient of hole flow is 0.65;

A_0 — orifice plate aperture is 17mm;

ρ_{t_1} — density at temperature of entering airflow (i.e., temperature at the flowmeter), kg/m³;

Δp — pressure difference on both ends of the orifice plate, kPa.

(2) Dimensionless equation of convective heat transfer（对流传热准数关联式）

There are many factors affecting the convective heat-transfer coefficient α, so it is very difficult to establish a general formula for various conditions. So far, the common approach is dimensional method to combine many influencing factors (physical quantities) into several number groups with dimension of 1 (dimensionless number), and then to determine the relationship between these parameters through experiments, that is, to obtain the correlation of α in different cases.

For a fluid by forced convection in the pipe, it is satisfactory to use a dimensionless equation as follow

$$Nu = A Re^m Pr^n \tag{2-42}$$

Where, Nusselt number $\quad Nu = \dfrac{\partial_i d_i}{\lambda}$

Reynolds number $\quad Re = \dfrac{d_i u \rho}{\mu} Nu$

Prandtl number $\quad Pr = \dfrac{C_p \mu}{\lambda}$

Physical parameters, such as λ, C_p, ρ, μ can be obtained according to qualitative temperature $\bar{t}$. For the heated airflow by forced convection in the pipe, Prandtl number (Pr) can be regarded as a constant, and n takes 0.4. Therefor the dimensionless equation is simplified as

$$Nu = A Re^m Pr^{0.4} \tag{2-43}$$

One content of this experiment is to plot the correlation graph of Re-$Nu/Pr^{0.4}$ on log-log coordinate paper, and induce the measured data into a dimensionless equation（在双对数坐标纸上画出 Re-$Nu/Pr^{0.4}$ 的关联图，并整理成准数方程）.

(3) Enhanced double-pipe exchanger and enhancement ratio（强化套管换热器及强化比）

Compared with ordinary heat exchangers, enhanced one has certain practical advantages: it can reduce the heat transfer area as well as the volume and weight of the heat exchanger, improve the heat transfer capacity of the heat exchanger, allow the heat exchanger work at lower temperature difference, and reduce heat exchanger power consumption by reducing the heat exchanger resistance. There are several approaches to enhance heat transfer. In this experiment, a spiral coil is inserted into the inner pipe of the heat exchanger to achieve the purpose of enhancement the heat transfer. The structure of the inserted spiral coil in the enhanced double-pipe exchanger is shown in Fig. 2-39.

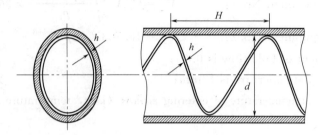

Fig. 2-39 Structure of the inserted spiral coil in the enhanced double-pipe exchanger
图 2-39 螺旋线圈强化管内部结构

The spiral coil fixed in the pipe is wound at a certain pitch with copper and steel wire below 3mm in inner diameter. In the region near pipe wall, the heat transfer is enhanced not only by rotating of fluid due to the action of the spiral coil, but also by periodical disturbance of the spiral wire of the coil. The strength of the fluid swirl is weak with a very fine spiral wire, so the resistance is small, which is beneficial to saving energy.

The main technical parameters of the spiral coil include the ratio of coil pitch H to inner diameter d of the pipe and the roughness of the pipe, and the ratio of length to diameter is an important factor affecting the heat transfer effect and the resistance coefficient.

Regardless of the influence of resistance, the enhancement ratio can be used as the evaluation index, namely Nu/Nu_0, where Nu and Nu_0 are the Nusselt number for enhanced pipe and ordinary pipe, respectively. The larger the ratio, the better the enhancement effect. The resistance coefficient is proportional to the heat transfer coefficient, which will inevitably lead to the decrease of heat transfer performance and the increase of energy consumption. Considering synthetically, the mode with large enhancement ratio and small resistance coefficient is perfect for enhancement mode（若不考虑阻力的影响，可用强化比作为评价指标，即 Nu/Nu_0，其中：Nu 是强化管的努塞特数，Nu_0 是普通管的努塞特数，其比值越大，强化效果越好。由于阻力系数随换热系数的增加而增大，从而导致换热性能的降低和能耗的增加。强化比大，阻力系数小的强化方式才是最佳的强化方式）。

III. Experimental device and process

(1) Schematic diagram of experimental equipment and flow chart

As shown in Fig. 2-40, the double-pipe exchanger consists of two tubes of different diameters concentrically interlocked together. The inner tube is made of copper, the outer tube is made of stainless steel, and both ends are fixed with stainless steel flanges. The air source is provided by a vortex air pump, airflow is adjusted by bypass regulating valve. After flowing the orifice meter, airflow enters the different branch selected by the branch control valve. The steam produced from the electric heating vessel rises naturally, and switches through different branches selected by the branch control valve into the tube side of measured exchanger. Finally, it ejected naturally from the stem vent at the other end.

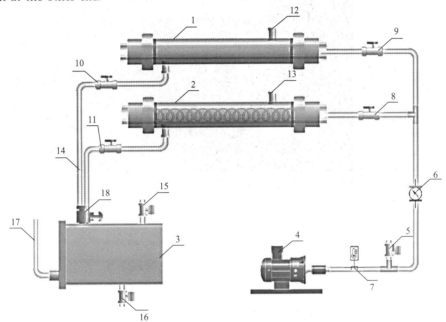

Fig. 2-40 Comprehensive experiment of convective heat transfer—
Device schematic diagram and flow chart

图 2-40 对流传热综合实验——实验装置示意图及流程

1—Ordinary double-pipe exchanger（普通套管换热器）；2—Enhanced double-pipe exchanger with spiral coils inserted（内插有螺旋线圈的强化套管换热器）；3—Electric steam boiler（电加热釜）；4—Vortex air pump（旋涡气泵）；5—Bypass regulating valve（旁路调节阀）；6—Orifice meter（孔板流量计）；7—Temperature test point of entering airflow［风机出口温度（冷流体入口温度）测试点］；8,9—Control valve on air branch（空气支路控制阀）；10,11—Control valve on steam branch（蒸汽支路控制阀）；12,13—Steam vent（蒸汽放空口）；14—Main road of rising steam（蒸汽上升主管路）；15—Water fill opening（加水口）；16—Drain opening（放水口）；17—Liquidometer（液位计）；18—Condensate reflux port（冷凝液回流口）

(2) Experimental simulation interface（Fig. 2-41, Fig. 2-42）

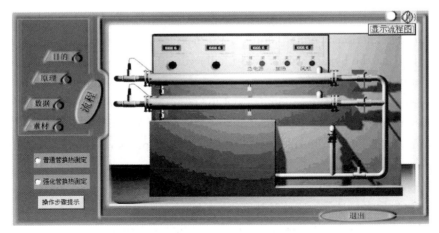

Fig. 2-41 Comprehensive experiment of convective heat transfer—Experimental simulation interface

图 2-41 对流传热综合实验——实验仿真界面

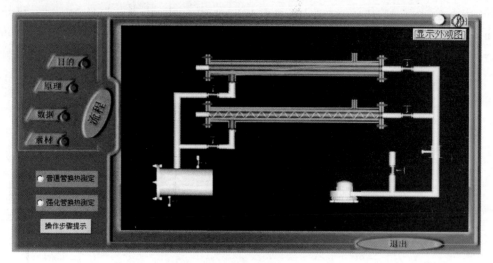

Fig. 2-42 Comprehensive experiment of convective heat transfer—Flow chart of experimental simulation
图 2-42 对流传热综合实验——实验仿真流程图

IV. Experimental steps

(1) Experimental operations

① Add water into the electric steam boiler to the red line at the upper end of the liquidometer.

② Add an appropriate amount of ice water to the thermos flask, and insert the cold end compensation couple into it（向冰水保温瓶中加入适量的冰水,并将冷端补偿电偶插入其中）.

③ Check the pipe valves and electrical switches. Set the electric heater switch to the heating state. Start the experiment after the steam discharge from the steam outlet（蒸汽排出口有蒸汽排出说明可以开始实验）.

④ Adjust the airflow by bypass valve, adjust 5-6 openings between minimum and maximum flowrate, and read each value after stabilization.

⑤ Change the branch, conduct the experiment in the enhanced double-pipe exchanger, and measure 5-6 groups of experimental data.

⑥ When the experiment ends, shut off the heater switch. Turn off the blower after 5 minutes, fully open the bypass valve, and finally cut off the power supply.

(2) Operating notes

① Since the electric thermocouple is used for temperature measurement, it is necessary to check the thermos for the presence of the ice-water coexistence mixture before the experiment, and check that the cold end of the thermocouple is completely immersed in the ice-water mixture.

② Check the water level in the steam boiler. If the water level is too low, replenish water in time（发现水位过低,应及时补充水）.

③ Ensure the unimpeded steam rising line, that is, one of the two steam branch control valves must be fully opened before heating the steam boiler（在给蒸汽加热釜加热之

Chapter 2 Experiment of Chemical Engineering Principle

前，两蒸汽支路控制阀之一必须全开）.

④ When switching the branch, open the required branch valve first, and then close the other side. It is important to open and close the control valve slowly to prevent pipe cutoff or sudden ejection due to excessive steam pressure（防止管线截断或蒸汽压力过大突然喷出）.

⑤ Ensure the unimpeded air line（空气管线）, that is, one of the two air branch control valves and the bypass control valve must be fully opened before connecting the fan power. When switching branch, turn off the fan power first, and then open or close the control valve.

⑥ After adjusting the flow rate, stabilize for at least 5-10 minutes, and then read the experimental data.

⑦ In the experiment, the amount of rising steam should remain stable, and steam continues to escape from the steam port.

V. Record and organize experimental data

（1） Experimental data recording（Fig. 2-43, Fig. 2-44）

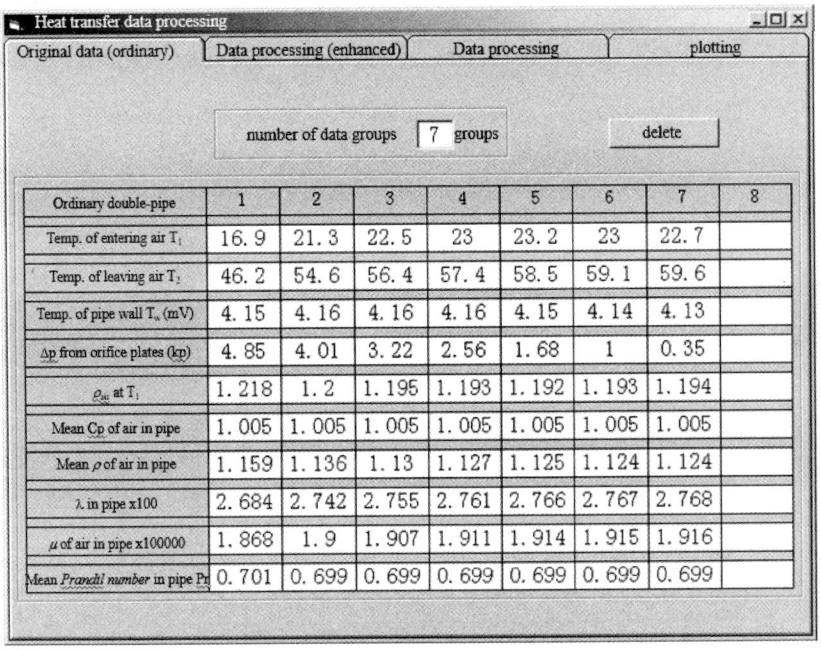

Ordinary double-pipe	1	2	3	4	5	6	7	8
Temp. of entering air T_1	16.9	21.3	22.5	23	23.2	23	22.7	
Temp. of leaving air T_2	46.2	54.6	56.4	57.4	58.5	59.1	59.6	
Temp. of pipe wall T_w (mV)	4.15	4.16	4.16	4.16	4.15	4.14	4.13	
Δp from orifice plates (kp)	4.85	4.01	3.22	2.56	1.68	1	0.35	
ρ_{air} at T_1	1.218	1.2	1.195	1.193	1.192	1.193	1.194	
Mean C_p of air in pipe	1.005	1.005	1.005	1.005	1.005	1.005	1.005	
Mean ρ of air in pipe	1.159	1.136	1.13	1.127	1.125	1.124	1.124	
λ in pipe ×100	2.684	2.742	2.755	2.761	2.766	2.767	2.768	
μ of air in pipe ×100000	1.868	1.9	1.907	1.911	1.914	1.915	1.916	
Mean Prandtl number in pipe Pr	0.701	0.699	0.699	0.699	0.699	0.699	0.699	

Fig. 2-43 Comprehensive experiment of convective heat transfer—
Data recording（ordinary double-pipe exchanger）
图 2-43 对流传热综合实验——数据记录（普通管）

（2） Experimental data collation（Fig. 2-45）

VI. Requirements for experimental report

① Pot the correlation graph of Re-$Nu/Pr^{0.4}$ on the log-log coordinate paper according to the experimental data using the ordinary and the enhanced double-pipe exchangers, and find the dimensionless equation（Fig. 2-46）.

Heat transfer data processing

Original data (ordinary) | Data processing (enhanced) | Data processing | plotting

number of data groups 6 groups　delete

Enhanced double-pipe	1	2	3	4	5	6	7	8
Temp. of entering air T_1	22.1	24	24.8	25.3	25.3	24.8		
Temp. of leaving air T_2	69.3	70.8	71.4	71.9	72.5	74.7		
Temp. of pipe wall T_w (mV)	3.91	3.91	3.92	4	3.93	3.9		
Δp from orifice plates (kp)	3.66	3.04	2.31	1.76	1.17	0.24		
ρ_{air} at T_1	1.197	1.189	1.186	1.184	1.184	1.186		
Mean Cp of air in pipe	1.005	1.005	1.005	1.005	1.005	1.005		
Mean ρ of air in pipe	1.108	1.102	1.1	1.098	1.097	1.094		
λ in pipe x100	2.8	2.812	2.817	2.82	2.822	2.828		
μ of air in pipe x100000	1.939	1.947	1.951	1.953	1.955	1.959		
Mean *Prandtl number* in pipe Pr	0.698	0.698	0.698	0.698	0.698	0.698		

Fig. 2-44 Comprehensive experiment of convective heat transfer—Data recording (enhanced double-pipe exchanger)

图 2-44 对流传热综合实验——数据记录（强化管）

Heat transfer data processing

Original data (ordinary) | Data processing (enhanced) | Data processing | plotting

processing　Return

Ordinary double-pipe	1	2	3	4	5	6	7	8
Temp. of pipe wall T_w (mV)	96.73	96.94	96.94	96.94	96.73	96.52	96.3	
Mean Temp. of air in pipe T	31.55	37.95	39.45	40.2	40.85	41.05	41.15	
ΔT between inlet and outlet air	29.3	33.3	33.9	34.4	35.3	36.1	36.9	
$\overline{\Delta t}$ between hot and could fluids	64.07	57.39	55.79	54.96	53.97	53.45	53.03	
Mean V_s in pipe (m^3/h)	45.45	41.87	37.63	33.6	27.27	21.06	12.47	
Mean u of air in pipe (m/s)	43.4	39.99	35.93	32.09	26.04	20.11	11.91	
Heat transfer rate Q (W)	430.86	442.21	402.41	363.69	302.35	238.55	144.39	
α ($W/m^2\cdot{}^\circ C$)	101.15	115.89	108.49	99.53	84.26	67.12	40.95	
Reynolds number Re	51835	46022	40987	36429	29466	22722	13448	
Nusselt number Nu	72.54	81.36	75.81	69.39	58.64	46.7	28.48	
$Nu/(pr^{0.4})$	83.62	93.89	87.48	80.08	67.67	53.89	32.86	
Enhanced double-pipe	1	2	3	4	5	6	7	8
Temp. of pipe wall T_w (mV)	91.63	91.63	91.84	93.54	92.05	91.41		
Mean Temp. of air in pipe T	45.7	47.4	48.1	48.6	48.9	49.75		
ΔT between inlet and outlet air	47.2	46.8	46.6	46.6	47.2	49.9		
$\overline{\Delta t}$ between hot and could fluids	41.55	39.74	39.23	40.58	38.44	36.09		
Mean V_s in pipe (m^3/h)	29.52	26.96	23.52	20.54	16.77	7.62		
Mean u of air in pipe (m/s)	28.19	25.75	22.46	19.62	16.01	7.28		
Heat transfer rate Q (W)	430.93	388.18	336.57	293.46	242.36	116.13		
α ($W/m^2\cdot{}^\circ C$)	155.98	146.92	129.03	108.78	94.83	48.4		
Reynolds number Re	31004	28051	24376	21232	17294	7822		
Nusselt number Nu	107.24	100.58	88.17	74.25	64.68	32.94		
Nusselt number Nu_o	57.74	53.48	48.02	43.19	36.9	20.08		
Reinforcement ratio Nu/Nu_o	1.86	1.88	1.84	1.72	1.75	1.64		
$Nu/(pr^{0.4})$	123.82	116.14	101.81	85.74	74.69	38.04		

Fig. 2-45 Comprehensive experiment of convective heat transfer—Data processing

图 2-45 对流传热综合实验——数据整理

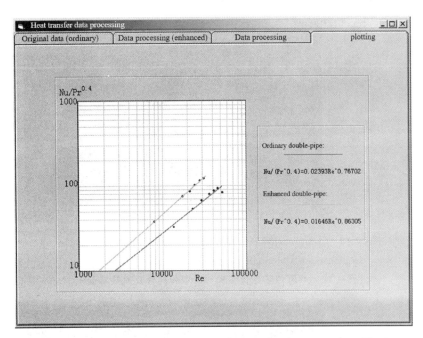

Fig. 2-46 Comprehensive experiment of convective heat transfer—Plotting

图 2-46 对流传热综合实验——数据作图

② According to the correlation plot of Re-$Nu/Pr^{0.4}$ obtained above, calculate the enhancement ratio Nu/Nu_0 of the enhanced to the ordinary double-pipe exchanger under the same flow rate.

③ Calculate the dimensionless equation by the least square method, and compared it with that by graphical method.

VII. Relevant materials (Fig. 2-47 to Fig. 2-52)

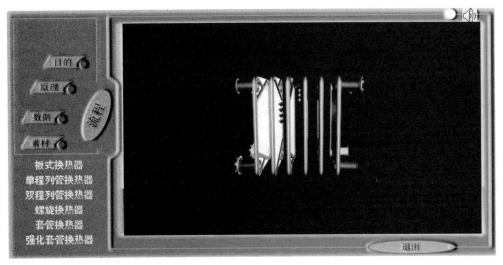

Fig. 2-47 Comprehensive experiment of convective heat transfer—
Demonstration of plate heat exchanger

图 2-47 对流传热综合实验——板式换热器演示

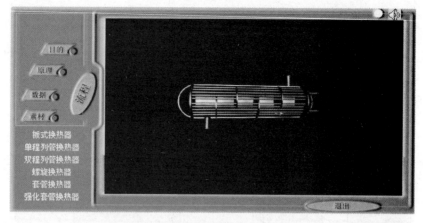

Fig. 2-48 Comprehensive experiment of convective heat transfer—
Demonstration of single-pass shell and tube heat exchanger

图 2-48 对流传热综合实验——单程列管式换热器演示

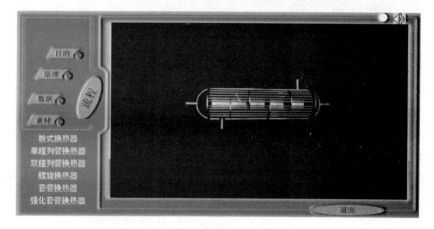

Fig. 2-49 Comprehensive experiment of convective heat transfer—
Demonstration of double-pass shell and tube heat exchanger

图 2-49 对流传热综合实验——双程列管式换热器演示

Fig. 2-50 Comprehensive experiment of convective heat transfer—
Demonstration of spiral heat exchanger

图 2-50 对流传热综合实验——螺旋换热器演示

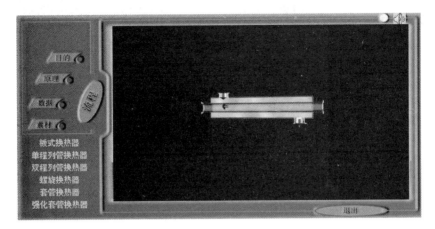

Fig. 2-51　Comprehensive experiment of convective heat transfer—
Demonstration of double-pipe exchanger

图 2-51　对流传热综合实验——套管换热器演示

Fig. 2-52　Comprehensive experiment of convective heat transfer—
Demonstration of enhanced double-pipe exchanger

图 2-52　对流传热综合实验——强化套管换热器演示

VIII. Questions

① How to correlate the dimensionless equation of convective heat transfer experimentally?

② What does the enhanced heat transfer process mean? What are the available approaches to enhance the heat transfer?

③ For gas-liquid convection heat transfer, how to improve the total heat transfer coefficient?

④ When the air velocity increases, will the temperature of air leaving the heat exchanger increase or decrease? Try to analyze the reason.

⑤ What is the impact of the flow direction of air and steam on the heat transfer effect in this experiment?

Experiment 6 Comprehensive experiment on the measurement of heat transfer coefficient K of heat exchanger
实验 6 换热器传热系数 K 值的测定综合实验

I. Experiment purpose

① Understand the structure characteristics and performance of double-pipe heat exchanger, tubular heat exchanger and spiral plate heat exchanger（套管式换热器、列管式换热器和螺旋板式换热器）.

② Measure the logarithmic average temperature difference（对数平均温度差）Δt_m in parallel flow（并流）and countercurrent flow（逆流）of the heat exchanger.

③ Determine the heat transfer coefficient K of the heat exchanger.

④ Determine the heat transfer performance of heat exchanger under different operating conditions.

II. Experimental principle

According to the basic heat transfer equation $Q = KA\Delta t_m$, the physical significance of total heat transfer coefficient K is the heat transferred through unit heat transfer area per unit time when the average heat transfer temperature difference is 1℃（当传热平均温度差为 1℃时，在单位时间内通过单位传热面积所传递的热量）. K value is an important parameter to measure the performance of heat exchanger. In the basic equation of heat transfer, the heat transfer Q is specified by the production task, and the temperature difference is related to the temperature of the cold and hot fluid at inlet and outlet of the heat exchanger, so the heat transfer area A is closely related to the total heat transfer coefficient K value. The reasonable determination of K value is an important issue in the calculation of heat exchanger.

For heat transfer to fluids without phase change in the heat exchanger, the constant specific heat of the fluid or the specific heat under the average temperature difference are assumed, the overall heat balance becomes:

$$Q = W_h C_{ph}(T_1 - T_2) = W_c C_{pc}(t_2 - t_1) \qquad (2\text{-}44)$$

Where, Q—heat load of heat exchanger, kJ/h or kW;

　　　　W—mass flow rate of fluid, kg/h;

　　　　C_p—average specific heat capacity of fluid, kJ/(kg·℃).

Subscripts c and h represent cold and hot fluids respectively, and subscripts 1 and 2 represent the inlet and outlet of the heat exchanger, respectively.

Suppose that, of the two temperature-difference at both ends of the heat exchanger, the larger value is Δt_2, and the smaller value is Δt_1, then the logarithmic mean temperature difference (LMTD) Δt_m can be obtained by the following formula:

$$\Delta t_m = \frac{\Delta t_2 - \Delta t_1}{\ln \dfrac{\Delta t_2}{\Delta t_1}} \tag{2-45}$$

Total heat transfer coefficient:
$$K = \frac{Q}{A \Delta t_m} \tag{2-46}$$

Where, A—heat transfer area of heat exchanger, m^2.

By changing the velocity / flow rate of cold/hot fluid at the inlet of the heat exchanger, the heat transfer performance of the heat exchanger under different conditions can be tested.

III. Experimental device and process

The flow chart of the comprehensive experimental device of hot-cold water heat exchanger is shown in Fig. 2-53. The heat exchanger is replaceable.

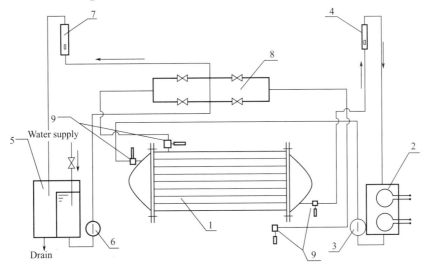

Fig. 2-53　Flow diagram of the experimental heat exchanger
图 2-53　换热器实验装置流程示意图

1—Heat exchanger（replaceable）［换热器（可更换）］；2—Heating water tank（加热水箱）；3—Hot water pump（热水泵）；4—Rotor flow meter（转子流量计）；5—Cold water tank（冷水箱）；6—Cold water pump（冷水泵）；7—Rotor flowmeter（转子流量计）；8—Parallel and countercurrent reversing valve set for cold flow（冷流并、逆流换向阀门组）；9—Temperature sensor（digital display temperature）［测温传感器（数显温度）］

The heater of the experimental device adopts technology of automatic temperature limit and automatic temperature control（自动限温和自动控温）. The inlet/outlet temperatures of cold/hot fluid is measured by temperature digital display instrument, and the temperature measuring point can be switched by keyboard switch.

Experimental device parameters

（1）Heat exchange area of heat exchanger（A）

① Double-pipe heat exchanger：$0.45m^2$

② Spiral plate heat exchanger：$0.65m^2$

③ Tubular heat exchanger：$1.05m^2$

(2) The total electric heating power: 7.5kW

(3) Hot water pump type: 12WGR-8

① Allowable working water temperature: < 80℃

② Rated flow: 8L/min

③ Rated lift: 8m

④ Motor: 220V, 90W

(4) Rotor flowmeter

① Model: LZB-10

② Measuring range: <160L/h

IV. Experimental steps and operating notes

(1) Experimental steps

① Preparation before experiment

a. Familiar with the working principle and performance of experimental devices and instruments.

b. Replace and install the heat exchanger to be tested.

c. According to parallel flow (or countercurrent) mode, open or close each valve of reversing valve group of the cold flow.

d. Fill the hot water tank with water.

② Experimental operation

a. Switch on the power, start the hot water pump (to improve the heating rate of hot water, do not open the cold-water valve first) and adjust the appropriate flow rate.

b. Adjust the temperature controller to a specified heating water temperature below 80℃.

c. Both the manual and automatic electric heaters of the heating water tank are powered.

d. Cut off the manual electric heater after the first action of the automatic voltage device, the heating system subsequently enters the automatic control state.

e. Select the key switch and temperature digital display instrument, observe and check the inlet/outlet temperature of cold/hot fluid of heat exchanger.

f. After the temperature of cold and hot fluid is stable, record the temperature values at temperature measuring points and the flow rate of both fluids on the flowmeter. To improve the accuracy of the experiment, measure the data every 5-10 minutes. Take the mean value of the 4 measuring data as the final data (取四次数据的平均值作为测定的数据).

g. Change the velocity/flow rate of cold/hot water at the inlet of the heat exchanger to determine the heat transfer performance under different operating conditions.

h. After the experiment, turn off the electric heater, then close the cold-water valve, and finally cut off the power supply.

(2) Operating notes

① The allowable heating temperature of hot fluid in the tank shall not exceed 80℃ (热流体在热水箱中加热温度不得超过80℃).

② After the experiment, drain all the water in the water tank.

V. Requirements for experimental report

① Determination the logarithmic mean temperature difference Δt_m of heat exchanger under parallel and countercurrent flow, respectively.

② Determine the total heat transfer coefficient K of the heat exchanger.

③ With heat transfer coefficient as ordinate and the flow rate of cold/water as abscissa, plot the heat transfer performance curve of heat exchanger on an ordinary coordinate paper.

VI. Questions

① Illustrate the structures and characteristics of double-pipe heat exchanger, tube heat exchanger and spiral plate heat exchanger.

② Compare the logarithmic mean temperature difference Δt_m of parallel and countercurrent flow.

③ When designing heat exchanger, how to determine the heat transfer coefficient K?

Experiment 7　Plate column distillation experiment
实验 7　板式塔精馏实验

I. Experiment purpose

① Familiar with the distillation process, master operation of the distillation experiment.

② Understand the structure of plate column; observe the vapor-liquid contact status on the tray.

③ Determine the overall efficiency and Murphree efficiency (plate efficiency) at total reflux ratio（全回流时的全塔效率及单板效率）.

④ Determine the overall efficiency at partial reflux ratio（部分回流时的全塔效率）.

⑤ Determine the concentration (or temperature) distribution of the whole column.

II. Experimental principle

In the distillation process, separation results from the multiple partial vaporization of reflux down-flowing from the top of the column or the multiple partial condensation of vapor rising from the still（塔釜）.

Reflux is a necessary condition for distillation operation（回流是精馏操作的必要条件）. The ratio of reflux to the overhead product is quantity called the reflux ratio（回流比）, which is the main parameter of distillation operation. Its value directly affects the separation effect and energy consumption of distillation operation. If the tower is operated at the minimum reflux ratio（最小回流比）, the number of plates needed for a given separation becomes infinite, which is not feasible in industry. At total reflux（全回流）, the number of plates is a minimum, but the rates of feed and both the overhead and bottom products are zero. Under this condition, the condensate at the top of the column entirely returns to the column, which is of no significance in the production. However, due to the minimum number of theoretical plates required at this time, it is easy to achieve stability. So it is often used in scientific research and industrial equipment parking and troubleshooting. Generally, the optimum reflux ration is 1.2-2.0R_{min} （通常回流比取最小回流比的 1.2~2.0 倍）.

(1) Plate efficiencies（塔板效率）

In a plate distillation column, mass transfer occurs when the vapor-liquid two phases contact each other on the trays. However, if the time of contact and the degree of mixing on the tray are insufficient, the efficiency of the stage or tray will not be 100%. The plate efficiency is usually used to indicate the degree of mass transfer on the tray.

The plate efficiency （塔板效率） is the main parameter to reflect the performance and operation condition of tray. There are many factors affecting the efficiency of the tray, which can be roughly summarized as: the physical properties of the fluid (such as viscosity, density, relative volatility and surface tension, etc.), the structure of the tray and the operating conditions, etc. As the factors affecting the plate efficiency are complex, by far it is still determined experimentally.

① The overall efficiency E_T（总板效率或全塔的效率）. It is simple to use but is the least fundamental, which reflects the average separation effect of each tray in the whole column. It is often referred in the design of plate column.

$$E_T = \frac{N_T - 1}{N_p} \times 100\% \tag{2-47}$$

Where, E_T—overall efficiency;

N_T—the number of theoretical plate;

N_p—the number of actual plate.

② Murphree efficiency (plate efficiency, 单板效率) E_M. It reflects the mass transfer effect on a single plate, and is an important data to evaluate the performance of tray, which is used in the studies on trays.

$$E_{ML} = \frac{X_{n-1} - X_n}{X_{n-1} - X_n^*} \tag{2-48}$$

Where, E_{ML}—Murphree efficiency (plate efficiency) expressed in liquid phase concentration;

X_n—actual concentration of liquid leaving the nth plate;

X_{n-1}—actual concentration of liquid entering the nth plate;

X_n^*—concentration of liquid in equilibrium with vapor leaving from the nth plate.

(2) Number of theoretical plates N_T（理论塔板数）

In the case of total reflux, both operating lines coincide with the diagonal in the x-y diagram （在全回流操作时，操作线与 x-y 图中的对角线相重合）. The minimum tray number required for a given separation may be found by constructing stages on an x-y diagram between compositions x_D and x_W, using the 45° line as the operating line for both sections of the column.

The number of theoretical plates at a certain reflux ratio can be determined by either plate-to-plate calculation（逐板计算法）or a graphical step-by-step construction（图解法）. The specific steps of graphic method are as follows：

① Plot the equilibrium curve（平衡曲线）of mixture to be separated on x-y diagram.

② Plot the operating line of rectifying section（精馏段操作线）according to the given overhead composition x_D and reflux ratio R. The operating line equation for rectifying section is：

$$y_{n+1} = \frac{R}{R+1} x_n + \frac{x_D}{R+1} \tag{2-49}$$

Where, y_{n+1}—the concentration (mole fraction) of vapor entering plate n in rectifying section;

x_n—the concentration (mole fraction) of liquid leaving plate n in rectifying section;

x_D—overhead composition (mole fraction) of distillate;

R—reflux ratio（回流比）.

$$R = \frac{L}{D} \tag{2-50}$$

Where, L—reflux stream in rectifying section, kmol/h;

D—overhead product, kmol/h.

③ Plot the q-line according to the thermal condition of the feed（进料热状态）, the feed line equation is：

$$y = \frac{q}{q-1} x - \frac{x_f}{q-1} \tag{2-51}$$

Where, x_f—concentration of the feed (mole fraction);

q—thermal condition of the feed（进料热状态）.

q is defined as the moles of liquid flow in the stripping section that result from the introduction of each mole of feed.

$$q = \frac{\text{heat required to vaporize per kilomole feed to saturated vapor}}{\text{kilomole latent heat of vaporization of feed}} = \frac{C_p(T_s - T_f) + r_c}{r_c} \tag{2-52}$$

Where, C_p—specific heat capacity of feed, kJ/(kmol·℃);

T_s—temperature of feed, ℃;

T_f—bubble point of feed, ℃;

r_c—latent heat of vaporization of feed, kJ/(kmol·℃).

④ Plot the stripping line（提馏段操作线）by connecting the point (x_w, x_w) on the diagonal and the intersection of q line with the operating line for the rectifying section.

⑤ Find the number of theoretical plates N_T by the graphical step-by-step construction.

III. Experimental device and process

(1) Schematic diagram of experimental equipment and flow chart

① Diagram and flow-chart of batch distillation (Fig. 2-54)

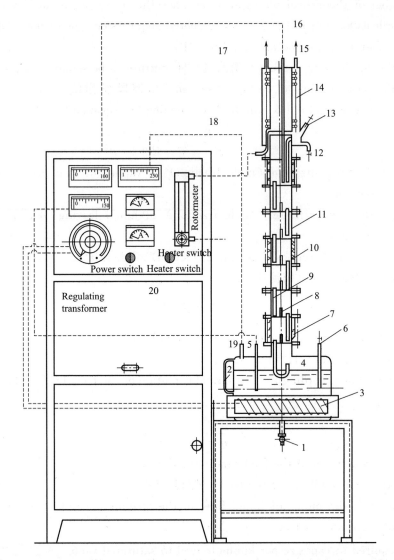

Fig. 2-54 Plate column distillation experiment——Diagram and flow chart of batch distillation
图 2-54 板式塔精馏实验——间歇精馏塔示意图及流程

1—Still sampling port（釜液取样口）；2—Level gage（液面计）；3—Boiler（加热器）；4—Still（塔釜）；5—Still temperature measuring nozzle（塔釜测温接管）；6—Feed inlet（加料口）；7—Glass tower section（玻璃塔节）；8—Overflow weir（溢流挡板）；9—Down-comer（降液管）；10—Tray（塔板）；11—Stainless steel tower section（不锈钢塔节）；12—Top sampling port（塔顶取样口）；13—Thermometer connection（温度计插孔）；14—Condenser（冷凝器）；15—Vent（放口接管）；16—Top temperature measuring nozzle（塔顶测温接管）；17—Cooling water outlet（冷却水出口）；18—Cooling water inlet（冷却水进口）；19—Still pressure gauge nozzle（测釜压接管）；20—Instrument cabinet（仪表柜）

The batch distillation column is composed of a small stainless steel sieve-tray column and an instrument cabinet. The horizontal still (220mm in diameter and 300mm in length) is equipped with level temperature measuring tube, piezo metric tube, feeding tube and still sampling port. Seven sieve-trays with a diameter of 50mm are arranged in the vertical column. The tray spacing is 100mm. The maximum power of tubular electric heater is

1kW. A copper coil condenser (ϕ10mm×1mm and length of 3.25m) is equipped at the top of the column. Water as coolant flows through inside the pipe while the vaporized steam condenses outside the pipe. The flow rate of water is measured by rotor flowmeter. The temperature of top and still of column is measured by platinum resistance thermometer and displayed by temperature indicator. The still pressure is displayed by pressure instrument.

② Schematic diagram and process of continuous distillation tower (Fig. 2-55)

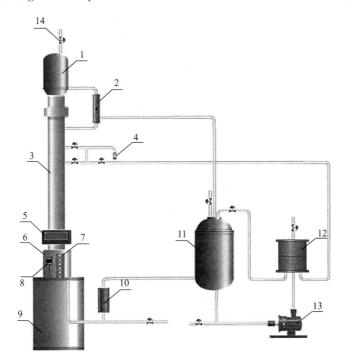

Fig. 2-55 Plate column distillation experiment—Diagram and flow chart of continuous distillation
图 2-55 板式塔精馏实验——连续精馏塔示意图及流程
1—Condenser (塔顶冷凝器); 2—Reflux distributor (回流比分配器); 3—Column (塔身); 4—Rotameter (转子流量计); 5—Viewing window (视盅); 6—Still (塔釜); 7—Still heater (塔釜加热器); 8—Temperature controlled heater (控温加热器); 9—Holder base (支座); 10—Cooler (冷却器); 11—Feed tank (原料液罐); 12—Buffer tank (缓冲罐); 13—Feed pump (进料泵); 14—Vent (放气阀)

a. Distillation column

The distillation column (ϕ57mm × 3.5mm) is equipped with a set of 10 sieve trays. Other pertinent detail are: 80mm tray spacing, 78.5mm^2 down-comer area, 12mm weir height, 6mm bottom gap. There are 43 small holes of 1.5mm diameter on each tray, arranged in an equilateral triangle with a spacing of 6mm. A viewing window is set on the column body to facilitate the observation of vapor-liquid contact on the tray. The sampling ports are arranged on the 1st-6th trays.

The distillation still with size of ϕ108mm×4mm×400mm is equipped with level gauge, electric heater (1.5kW), temperature controlled electric heater (200W), temperature-measuring nozzle, pressure measuring port and sampling port. The still can be treated as a theoretical plat. A coil condenser with the heat exchange area of 0.06m^2 is equipped at the top of the column. Water as coolant flows through inside the pipe while the vaporized steam condenses outside the pipe.

b. Reflux distribution device（回流分配装置）

The reflux distribution device consists of a reflux distributor and a controller, which is composed of a control instrument and a solenoid coil（回流分配装置包括一个回流分配器和一个由控制仪器和电磁线圈组成的控制器）. The glass reflux distributor consists of an inlet tube, two outlet tubes and a drainage rod. The two outlet tubes are used for reflux and extraction respectively. The drainage rod is a glass rod of ϕ 4mm with an iron core inside. The condensate from overhead condenser flows down the drainage rod. The reflux or extraction operation of the condenser on top of the tower is completed under the control of controller. That is, when the controller circuit is connected, the electromagnetic coil will suck up the drainage bar, the operation is in the production status of this time. When the controller circuit is broken, the electromagnetic coil does not work, the drainage rod naturally droops, and the operation is in the reflux status. The reflux distributor can be controlled manually by a controller or automatically by a computer.

c. Measurement and control system（测控系统）

In this experiment, the parameters (including overhead temperature, still temperature, column-tracing temperature, still heating temperature, whole tower pressure drop, heating voltage, feed temperature and reflux ratio) are measured by artificial intelligence instrument. The introduction of this system not only makes the experiment more simple and fast, but also realizes the computer online data acquisition and control.

(2) Experimental simulation interface (Fig. 2-56)

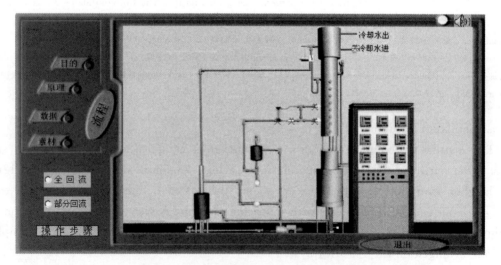

Fig. 2-56　Plate column distillation experiment—Experimental simulation interface
图 2-56　板式塔精馏实验——实验仿真界面

IV. Experimental steps

(1) Experimental operations

① Batch distillation operation（间歇精馏塔操作）

a. Before operation, feed ethanol of 15% mass concentration into the still until the liquid in still to 2/3 of the liquid level gauge.

b. Switch on the electric heating, adjust the voltage and current intensity (the normal operating current range is 3-4A).

c. Observe the temperature change at top and bottom column. When the rising vapor or reflux appears on the first tray of column, open the cooling water to sustain all vapor condensed.

d. Keep the total reflux for a period of time until the normal vapor-liquid bubbling and constant temperature of both top and bottom column. Sample at top and still of column respectively, and determine the composition of sample by a liquid specific gravity balance. See Appendix 3 for the use of liquid gravity balance.

e. At the end of the experiment, slowly reduce the heating current to zero, cut off the power. When there is no reflux in the column, close the cooling water valve.

② Continuous distillation operation（连续精馏塔操作）

a. Familiar with the distillation process referring to the flow chart, and make clear the function of each button on the instrument cabinet.

b. For total reflux operation, prepare the mixture of ethanol (20%-25% in molar fraction) and n-propanol in the material storage tank. Start the pump to feed mixture into column until the liquid height in still reaches 250-300mm.

c. Start the still heater and column tracing. Check the temperature of still, overhead and column body respectively. Observe the vapor-liquid contact status on the tray through viewing window. Open the cooling water control valve of the top condenser when feed liquid appears on the tray.

d. Determine the single plate efficiency and overall efficiency in the case of total reflux. After the column operation are stable, take samples at the overhead, still and two adjacent trays. Determine the samples with Abbe refractometer (repeat 2-3 times), and record the operation parameters.

e. After the total reflux operation is stable, adjust the feed concentration according to the concentration on the tray. Open the feed pump, set the feed amount and reflux ratio, measure the total efficiency under partial reflux condition. The suggestive amount is 30-50mL/min, reflux ratio is 3-5. Keep the liquid level in still constant (adjust the discharge of the residue in still). Be sure to open the cooling water control valve of still before discharging the still residue. After the operation is stable, sample at top and still of column respectively, and determine the composition of sample by a Abbe refractometer. The usage of Abbe refractometer is shown in Appendix 4.

f. At the end of the experiment, stop feeding, close the still heater and column tracing. After there is no feed liquid in the viewing window, cut off the water supply of overhead condenser and still cooler, cut off the power and clean up the site.

(2) Operating notes

① Adjust the voltage and current intensity to control the electric heating. The normal bubbling state of vapor-liquid is appropriate（以塔板汽液两相正常鼓泡为宜）. If the electric heating is too heavy, it is easy to cause liquid steam and destroy the normal operation of distillation.

② The amount of the cooling water should be appropriate to save water. However, if the cooling water is insufficient, the reflux temperature is too low, resulting in vapor ejection from the top vent during atmospheric distillation operation.

③ When sampling from batch distillation still, open sampling plug slowly to avoid scalding.

V. Record and organize experimental data（Fig. 2-57 to Fig. 2-63）

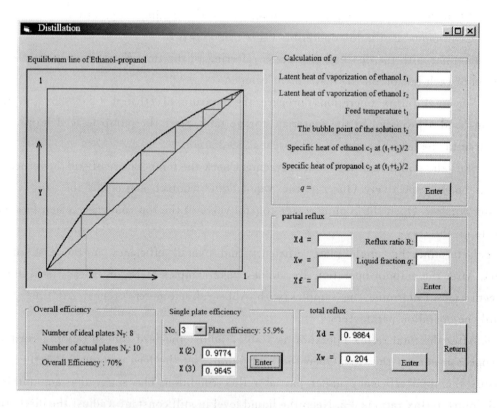

Fig. 2-57　Plate column distillation experiment——The overall efficiency and single plate efficiency under total reflux ratio（ethanol-propanol）

图 2-57　板式塔精馏实验——全回流条件下的总板效率和单板效率（乙醇-丙醇）

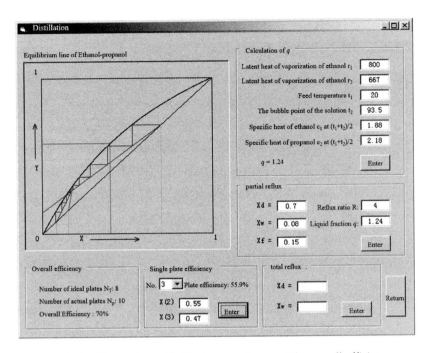

Fig. 2-58 Plate column distillation experiment—The overall efficiency and single plate efficiency under partial reflux ratio (ethanol-propanol)

图 2-58 板式塔精馏实验——部分回流条件下的总板效率和单板效率（乙醇-丙醇）

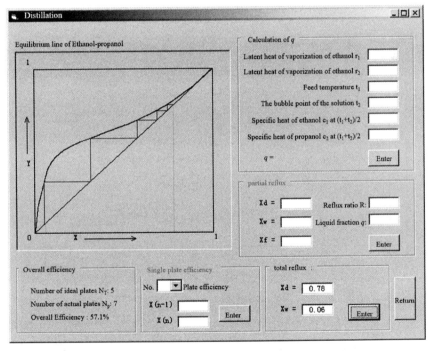

Fig. 2-59 Plate column distillation experiment—The overall efficiency under total reflux ratio (ethanol-water)

图 2-59 板式塔精馏实验——全回流条件下的总板效率（乙醇-水）

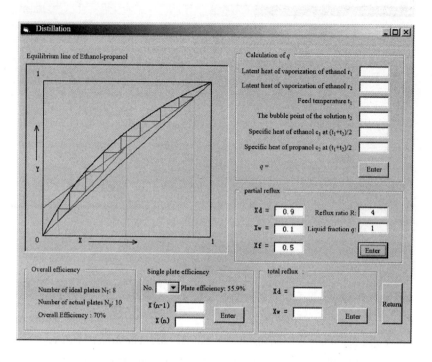

Fig. 2-60　Plate column distillation experiment—Feed saturated liquid（ethanol-propanol）
图 2-60　板式塔精馏实验——饱和液体进料（乙醇-丙醇）

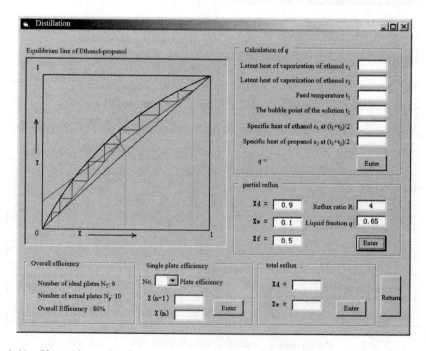

Fig. 2-61　Plate column distillation experiment—Feed partially vaporized（ethanol-propanol）
图 2-61　板式塔精馏实验——气液混合物进料（乙醇-丙醇）

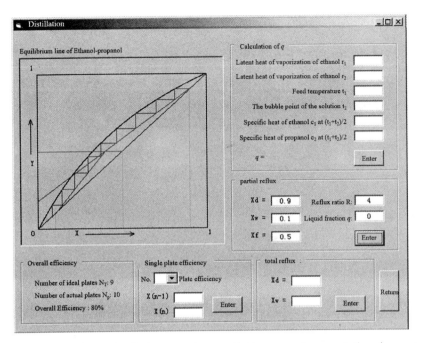

Fig. 2-62 Plate column distillation experiment—Feed saturated vapor (ethanol-propanol)
图 2-62 板式塔精馏实验——饱和蒸气进料（乙醇-丙醇）

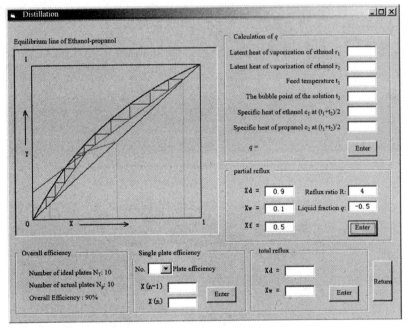

Fig. 2-63 Plate column distillation experiment—Feed superheated vapor (ethanol-propanol)
图 2-63 板式塔精馏实验——过热蒸气进料（乙醇-丙醇）

VI. Requirements for experimental report

(1) Plot x-y equilibrium diagram on an ordinary graph paper, calculate the theoretical plate number by graphical step-by-step construction method（在普通坐标纸上绘制 x-y 相平衡图，用图解法求出理论板数）.

(2) Calculate the overall efficiency and single plate efficiency（求出全塔效率和单板效率）.

VII. Relevant materials（Fig. 2-64 to Fig. 2-68）

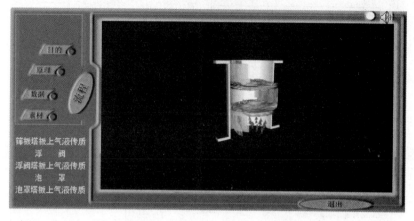

Fig. 2-64　Plate column distillation experiment—Demonstration of vapor-liquid mass transfer on sieve tray
图 2-64　板式塔精馏实验——筛板塔板上气液传质演示

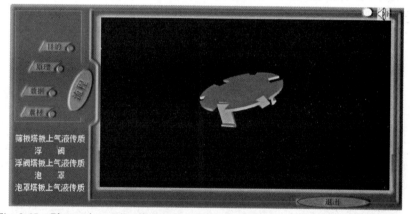

Fig. 2-65　Plate column distillation experiment—Demonstration of floating valve tray
图 2-65　板式塔精馏实验——浮阀塔板演示

Fig. 2-66　Plate column distillation experiment—Demonstration of vapor-liquid mass transfer on floating valve tray
图 2-66　板式塔精馏实验——浮阀塔板上气液传质演示

Fig. 2-67   Plate column distillation experiment—Demonstration of bubble cap
图 2-67   板式塔精馏实验——泡罩演示

Fig. 2-68   Plate column distillation experiment—Demonstration of vapor-liquid mass transfer on bubble cap tray
图 2-68   板式塔精馏实验——泡罩塔板上气液传质演示

VIII. Questions

① What are the characteristics of total reflux operation and its practical significance in production?

② How to keep the normal operation of the distillation column? What is the effect of too much or too little heating current on operation?

③ At the reflux ratio of $R < R_{min}$, is the distillation still operable?

④ To separating the ethanol-water solution, whether anhydrous alcohol can be obtained by increasing the height of the column appropriately?

⑤ How to determine the feed plate location?

⑥ How to operate the distillation column at atmospheric pressure? If necessary, how to perform vacuum operation?

Experiment 8 Packed column distillation experiment
实验 8 填料精馏塔实验

I. Experiment purpose

① Observe the vapor-liquid flow status during the distillation process in a packed distillation column.

② Master the method of measuring the height equivalent of theoretical plate（HETP）（等板高度）.

③ Understand the effect of reflux ratio on distillation operation.

II. Experimental principle

The distillation column is a vapor-liquid mass transfer equipment to realize the separation of liquid mixture. The distillation column can be divided into plate column（板式塔）and packed column（填料塔）. In the plate column, the vapor-liquid pair countercurrent contact plate-by-plate. While packed column is used for continuous differential countercurrent contacting of gas and liquid along the height of the packing layer. Packing, divided into dumped packing（散装填料）and structured packing（规整填料）, is the main component of the packed column. Dumped packing include Raschig ring, Pall ring, Cascade Mini-Ring, Berl saddle, Intalox saddle, θ-ring and so forth; Structured packing mainly refer to Mellapak packing and Metal Gauze Packing.

Because of the complexity of the vapor-liquid two-phase mass transfer process in the packed column, there are many influencing factors, including packing characteristics, gas-liquid contact condition and the physical properties of the two-phase. Calculations of the packing height under certain separation task, therefore, rest on direct experimental data or the empirical formula with the similar packing type, operation condition and separation system.

Two methods are commonly used to determine the packing height.

(1) Transfer unit method（传质单元法）

Because the mass transfer mechanism of packed column is that the composition of gas-liquid changes continuously along the packing layer rather than the step change, transfer unit method is the most suitable and widely used in the design calculation of packed column for absorption, desorption and extraction operation. The equation for column height can be written as follows:

Packing height=height of a transfer unit（HTU）×number of transfer units（NTU）

（填料层高度＝传质单元高度×传质单元数）

$$Z = H_{OL} N_{OL} = \frac{L}{K_X a \Omega} \int_{X_2}^{X_1} \frac{\mathrm{d}X}{X^* - X} \tag{2-53}$$

or

$$Z = H_{OG} N_{OG} = \frac{V}{K_Y a \Omega} \int_{Y_2}^{Y_1} \frac{\mathrm{d}Y}{Y^* - Y} \tag{2-54}$$

(2) Method of HETP (the height equivalent of theoretical plate) (等板高度法)

In the distillation calculation, the theoretical plate number (理论板数) is generally used to express the separation effect, so the height of the packed tower is then

$$Z = \text{HETP} \times N_T \tag{2-55}$$

Where, Z—packing height, m;

N_T—the number of theoretical plates;

HETP—the equivalent height of theoretical plate, m.

HETP means that the separation effect is equivalent to the packing layer height of a theoretical plate, also known as equivalent height (当量高度), in unit of m. In the packing tower design, if the HETP of the selected packing cannot be found, it can be directly determined by experiment.

For the mixture of binary components under the total reflux, the composition sampled at top and bottom of column are measured respectively after the distillation process is stable. The number of theoretical plate can be obtained with Fenske equation (芬斯克方程) or by graphical construction method in x-y diagram.

Equation (2-56) is the Fenske equation:

$$N_{\min} + 1 = \frac{\lg\left[\left(\dfrac{x_A}{x_B}\right)_D \left(\dfrac{x_B}{x_A}\right)_w\right]}{\lg \bar{\alpha}} \tag{2-56}$$

Where, $N_{\min}$—number of theoretical plates at total reflux;

$\left(\dfrac{x_A}{x_B}\right)_D$—molar ratio of volatile component to less volatile component at the top of column;

$\left(\dfrac{x_B}{x_A}\right)_w$—molar ratio of less volatile component to volatile component at the bottom of column;

$\bar{\alpha}$—geometric mean relative volatility of total column, $\bar{\alpha} = \sqrt{\alpha_{\text{top}} \cdot \alpha_{\text{bottom}}}$.

In partial reflux distillation (部分回流的精馏操作), the number of theoretical plates can be obtained from Fenske equation, Gilliland correlation (吉利兰图) or by graphical construction method in x-y diagram.

With determined theoretical plate number N_T and actual measured packing height Z,

$$\text{HETP} = Z/N_T \tag{2-57}$$

III. Experimental device and process (Fig. 2-69)

The packed distillation column is composed of two parts: a 1.2m high glass column body with diameter of 25mm and an instrument control cabinet. Stainless steel wire mesh θ-ring packing (2.5mm × 2.5mm) are dumped at random into the column. The length of rectifying-and stripping-section can be selected according to requirement. The still volume is 1000mL. The power of electric heating sleeve is 100—500W controlled by adjustable current meter and AI intelligent instrument. The top and bottom temperature is displayed by platinum resistance sensor and high precision digital display instrument. A reflux splitter is equipped next to the top condenser to control the amount of reflux and overhead product by time relay (reflux ratio controller) in the range of 1∶99-99∶1.

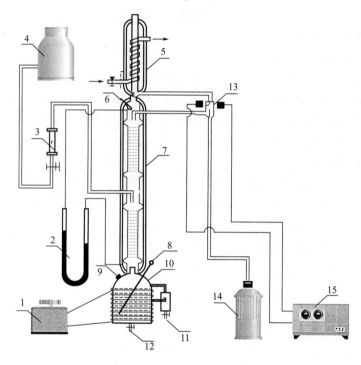

Fig. 2-69 Packed column distillation experiment—Device schematic diagram and flow chart
图 2-69 填料精馏塔实验——实验装置示意图及流程

1—Voltage regulator（调压器）; 2—U-tube gauge（U 形管压差计）; 3—Rotameter（转子流量计）; 4—Feed tank（料液罐）; 5—Condenser（冷凝器）; 6—Top thermometer（塔顶温度计）; 7—Packed column（填料精馏塔塔体）; 8—Bottom thermometer（塔釜温度计）; 9,11—Still sampling port（塔釜取样口）; 10—Still（塔釜）; 12—Vent（塔釜放空口）; 13—Reflux splitter（回流分配器）; 14—Product tank（产品罐）; 15—Reflux controller（回流控制仪）

IV. Experimental steps

(1) Experimental operations

① Add the prepared feed mixture of n-heptane-methyl cyclohexane（正庚烷-甲基环己烷）into the still（塔釜）until the liquid level exceeds the position of platinum resistance.

② Control the heating current within 0.1-1.5A by adjusting the knob, and turn on the cooling water at the same time.

③ When the still liquid begins to boil, turn on the insulation switch of column body and adjust the regulating knob to maintain the current at 0.1-0.3A. Observe the liquid state in the column to avoid channeling. After heating up, observe the temperature change at both top and bottom column. When the top steam begins to condense, carry out the total reflux operation（全回流操作）.

④ Keep the total reflux for a certain time until the temperature of top and still is stable. Sample at top and still respectively, and determine the refractive index of n-heptane-methyl cyclohexane at 25℃ by an Abbe refractometer（阿贝折射仪）. The composition of the sample can be derived from the composition-refractive index relationship（see Appendix 4.2）.

⑤ Turn on the reflux ratio switch to change the reflux ratio within 1∶6-1∶2. Carry out the partial reflux operation following the same operation as Step ④.

⑥ At the end of the experiment, firstly adjust the insulation current and the still heating current to zero, and then turn off the reflux ratio controller. After the top-and still-temperature cooled to room temperature, turn off the still heating and insulation power, and finally turn off the main power and cooling water.

（2）Operating notes

① To achieve an effective mass transfer area, ensure that the packing surface is wetted by the flowing liquid during the experiment.

② Released the retention liquid in the pipeline first during sampling to ensure the accuracy of measurement.

V. Requirements for experimental report

（1）Calculate the number of theoretical plates under total reflux and partial reflux（按全回流和部分回流两种情况计算理论板数）.

（2）Calculate the HETP of packing, and plot the relation curve between reflux ratio and HETP（计算填料等板高度，并做出回流比与等板高度的关系图）.

VI. Questions

① How to measure the HETP of packing by experiment directly? What is the significance of measuring HETP?

② What is the relationship between packing wettability and mass transfer efficiency? How to ensure the wettability of the packing during the experiment?

Experiment 9　Determination of fluid mechanics of plate column
实验9　板式塔流体力学性能测定

I. Experiment purpose

① Observe the behavior of vapor and liquid flowing on the tray.

② Determine the relationship of pressure drop with superficial gas velocity, froth entrainment rate with superficial gas velocity, and weeping rate with superficial gas velocity, respectively.

③ Study the influence factors to capacity performance chart of plate column, and plot the sieve-tray capacity performance chart.

II. Experimental principle

A sieve plate is designed as vapor-liquid mass transfer equipment in step-by-step contact

（板式塔为逐级接触的气-液传质设备）. The liquid flows across the plate and passes over a weir（堰）to a downcomer（降液管）leading to the plate below. The rising stream is longitudinally through the sieve hole, the floating valve, or bubble-cap slot in the tray, and intimate contact with liquid layer. The flow pattern on each plate is therefore crossflow rather than countercurrent flow. The mass transfer efficiency of a tray depends largely on the hydrodynamic status on the tray.

(1) Contact status of vapor-liquid on the tray and abnormal flow on the tray（塔板上的气-液两相接触状况及不正常的流动现象）

① Three contact states of vapor-liquid phases on the tray

a. At low vapor rates, the vapor-liquid two phases are in bubbling contact state（鼓泡接触状态）. An obvious clear liquid layer accumulates on the tray, and the vapor in the form of bubbles is dispersed in the clear liquid. The mass transfer between vapor-liquid two phases occurs on the bubble surface（气液两相在气泡表面进行传质）.

b. At higher vapor velocity, the vapor-liquid two phases are in foam contact state（泡沫接触状态）. The supernatant liquid on the tray becomes obviously thinner and is visible only on the surface of the tray. The liquid layer decreases with the increase of gas velocity accompanied by a large amount of foam on the tray. Liquid mainly exists between very dense bubbles in the form of constantly renewed liquid film. The mass transfer between gas-liquid two phases occurs on the surface of liquid film（气液两相以液膜表面进行传质）.

c. At very high vapor velocity, the vapor-liquid two phases are in spray contact state（喷射接触状态）. The liquid disperses among vapor in the form of continuously renewed droplets. The mass transfer between gas-liquid two phases occurs on the surface of liquid drops（气液两相以液滴表面进行传质）.

② Abnormal flow on the tray（塔板上不正常的流动现象）

a. Weeping（漏液） Sieve trays rely on vapor velocity to exclude liquid from falling through the perforations in the tray floor. If the vapor velocity is much lower than design, the dynamic pressure is insufficient to prevent liquid from flowing down through some of the holes, this condition is called weeping.

b. Froth entrainment（雾沫夹带） As the rising vapor passes through the liquid layer on the tray, the liquid droplets on the tray is brought back up to the tray above resulting in concentration back mixing.

c. Flooding（液泛） When the pressure drop on a tray is too high, the flow on that tray is inhibited. At this point, the liquid level increases and the foam height equals the tray spacing, leading to a large carryover of liquid to the tray above.

(2) Determination of fluid mechanics（流体力学性能测定）

① Pressure drop（压降） Two pressure taps, connected to the two ends of the U-tube manometer respectively, are set above and below the tray to determine the pressure drop of the vapor passing through the tray.

Pressure drop usually includes dry plate pressure drop（干板压降）and liquid layer pressure drop（液层压降）. When only the vapor passes through the tray, the dry plate pressure drop is mainly due to resistance through the sieve hole. On the other hand, as the vapor passes through the transparent liquid and foam layers on the tray, it overcomes resistance and creates a liquid layer pressure drop.

② Entrainment rate（雾沫夹带率） A liquid collecting plate is set on the tray to collect the liquid brought up by the vapor from the lower tray. Then the collected liquid is drawn out through a conduit. Division of vapor rate gives for the entrainment rate:

$$\text{Entrainment rate } e = \frac{\text{flow rate of entrained liquid}}{\text{flow rate of vapor}} \times 100\% \qquad (2\text{-}58)$$

③ Weeping rate（泄漏率） A liquid outlet is set under the tray to collect and measure the liquid flowing from the sieve hole. The weeping rate is given by the equation:

$$\text{Weeping rate } Q = \frac{\text{flow rate of weeping}}{\text{flow rate of liquid}} \times 100\% \qquad (2\text{-}59)$$

(3) Fluid dynamics model of sieve plate（筛板的流体力学模型）

① Pressure drop

$$\Delta p = \Delta p_c + \Delta p_L \qquad (2\text{-}60)$$

Where, Δp—total pressure drop, Pa;
Δp_c—pressure drop of dry tray, Pa;
Δp_L—pressure drop of liquid layer on the tray, Pa.

$$\Delta p_c = 0.051 \rho_v g \left(\frac{u_o}{C_o}\right)^2 \qquad (2\text{-}61)$$

Where, ρ_v—vapor density, kg/m³;
g—acceleration of gravity, m/s²;
u_o—sieve vapor velocity, m/s;
C_o—sieve hole flow coefficient, obtained from Fig. 2-70.

The pressure drop on the sieve plate due to the height of the liquid layer Δp_L, namely the effective resistance of the liquid layer h_L:

$$\Delta p_L = \rho_L g h_i \qquad (2\text{-}62)$$

Where, ρ_L—liquid density, kg/m³;
g—acceleration of gravity, m/s²;
h_i—effective resistance of liquid layer, m, obtained from Fig. 2-71.

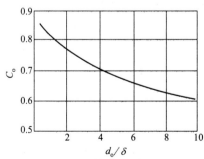

Fig. 2-70 Flow coefficient of dry sieve pore

图 2-70 干筛孔的流量系数

δ—thickness of tray（板厚）; d_o—pore diameter（孔径）, mm

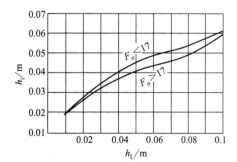

Fig. 2-71 Effective liquid layer resistance

图 2-71 有效液层阻力

In the diagram, the kinetic energy factor of vapor is expressed as:
$$F_o = u_o \sqrt{\rho_v}$$
Equivalent level of clear liquid on the tray is:
$$h_L = h_w + h_{ow}$$
Where, h_w—weir height, m;

h_{ow}—height of liquid flow on the weir, m.

Effective resistance h_i, can be calculated by the following equations:
When $F_o < 17$,
$$h_i = 0.005352 + 1.4776 h_L - 18.6 h_L^2 + 93.54 h_L^3 \tag{2-63}$$
When $F_o > 17$,
$$h_i = 0.006675 + 1.2419 h_L - 15.64 h_L^2 + 83.45 h_L^3 \tag{2-64}$$

② Entrainment amount (雾沫夹带量)
$$e_v = \frac{5.7 \times 10^{-5}}{\sigma} \left(\frac{u_G}{H_T - h_f}\right)^{3.2} \tag{2-65}$$

Where, e_v—entrainment amount;

σ—surface tension of liquid, N/m;

u_G—gas velocity calculated based on effective area, m/s;

H_T—plate spacing, m;

h_f—height of bubbling layer on the tray, m.

$$h_f = \frac{h_L}{\phi} \tag{2-66}$$

Where, ϕ is the mean relative density of the bubbling liquid, normally, $\phi = 0.4$.
Then,
$$h_f = 2.5 h_L \tag{2-67}$$
$$u_G = \frac{V_s}{A_T - A_f} \tag{2-68}$$

Where, V_s—vapor flow rate, m³/s;

A_T—cross-sectional area of the column, m²;

A_f—cross-sectional area of the down-comer pipe, m².

③ Weeping In order to avoid weeping of the sieve plate, the lower limit of vapor flow rate u_{ow} can be calculated according to the following empirical formula:
$$u_{ow} = 4.4 C_o \sqrt{\frac{(0.0056 + 0.13 h_L - h_\sigma) \rho_L}{\rho_v}} \tag{2-69}$$

Where, u_{ow}—vapor velocity through the sieve hole, m/s;

C_o—sieve flow coefficient;

h_L—height of clear liquid layer on the tray, m;

h_σ—equivalent height of the liquid column to the surface tension of the liquid, m.

$$h_\sigma = \frac{4\sigma}{d_o \rho_L g} \tag{2-70}$$

Where, σ—liquid surface tension, N/m;

g — acceleration of gravity, m/s^2;
d_o — hole diameter, m;
ρ_L — liquid density, kg/m^3.

(4) Capacity performance chart of plate and turndown ratio (塔板负荷性能图及操作弹性)

For certain vapor-liquid system, there exists an area of satisfactory operation for vapor and liquid loading within a plate column. A typical capacity chart is shown in Fig. 2-72. Each line represents a limit operating condition: Line 1 is drawn from the percentage flooding limit which stands for excessive froth entrainment; Curve 2 from the operating limit equation of downcomer flooding; Line 3 is the downcomer back-up limitation; Line 4 is the lower limit of the vapor flow which is set by the condition of excessive weeping; And line 5 is the lower limit of volumetric liquid flow rate. The capacity performance chart can be determined experimentally.

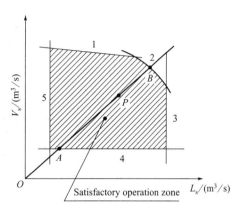

Fig. 2-72 Capacity performance chart
图 2-72 塔板负荷性能图

Turndown ratio (操作弹性) is an important index to evaluate the performance of a column. Under the operating liquid-vapor ratio, as shown in Fig. 2-72, the ratio of the maximum vapor load V_{max} to the allowable minimum vapor load V_{min} at the intersection of the operation line OAB and the boundary curve is called turndown ratio.

$$\text{turndown ratio} = \frac{V_{max}}{V_{min}} \tag{2-71}$$

When designing the tray, the structural parameters of the tray can be properly adjusted to make the operating point P at a moderate position in the diagram to improve the turndown ratio of the column.

III. Experimental device and process (Fig. 2-73)

The plexiglass column body for measuring fluid mechanics of plate column is divided into three sections connected by flanges. The upper one is a foam entrainment collecting tray, the middle part is a replaceable experimental tray (sieve tray, floating valve tray, bubble cover tray), and the bottom tray is a weeping rate measuring tray. The vapor enters the column from below and flows out from the top; the liquid is fed from the middle tray and flows out along the downcomer of the lower tray.

Determination of pressure drop of tray (塔板压降的测定): Two pressure taps, connected to the two ends of the U-tube manometer respectively, are set above and below the tray. The reading of the pressure difference meter directly reflects the pressure drop of the tray.

Determination of entrainment (雾沫夹带的测定): drain out the liquid under the entrainment collecting plate and simultaneously measured the flow rate.

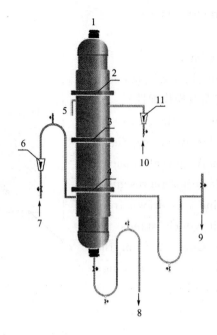

Fig. 2-73 Determination of fluid mechanics of plate column—Device schematic diagram and flow chart
图 2-73 板式塔流体力学性能测定——实验装置流程示意图

1—Air outlet（空气出口）；2—Entrainment collection plate（雾沫夹带收集板）；3—Experiment tray（sieve plate，float valve，bubble cap tray）[实验塔板（筛板、浮阀、泡罩塔板）]；4—Leakage rate measurement plate（泄漏率测定板）；5—Foam entrainment liquid outlet（雾沫夹带液出口）；6—Intake flow meter（进气流量计）；7—Air entry（空气进口）；8—Leakage liquid outlet（泄漏液出口）；9—Water outlet（出水口）；10—Water intake（进水口）；11—Inlet flow meter（进水流量计）

Determination of weeping rate（泄漏率的测定）：drain out the weeping liquid at bottom of the column. Calculate the weeping rate by measuring the leakage velocity after collection.

Main structural parameters of tray：

Column inner diameter $D=190$mm；Plate spacing $H_T=200$mm；

Length of exit weir $L_W=100$mm；Exit weir height $H_W=25$mm；

Diameter of downcomer $D=21$mm；

Height of the bottom gap of the lowering pipe $H_0=18$mm；

Sieve hole diameter $D_0=3$mm；Number of sieve $n=210$.

Other structural dimensions，such as plate thickness σ，should be measured according to the structure of the tray.

IV. Experimental steps

(1) Experimental operations

① Be familiar with the process of experimental setup and understand the function of each part.

② Start the fan to change the vapor flow rate，measure the relationship between dry plate pressure drop and vapor velocity.

③ Change the vapor flow rate at a selected liquid flow rate，measure the tray pressure drop, foam entrainment rate and weeping rate.

④ Change the flow rate of the liquid, and repeat Step ③.

⑤ At the end of the experiment, turn off the water first, and then turn off the gas.

(2) Operating notes

① When testing the relationship between pressure drop and vapor velocity under a certain liquid load, the water should be supplied first and the gas opened later（在做一定液体负荷下的压降与气速关系实验时，应先通水，后开气）.

② If the liquid flow is set, gas quantity should not be too small to lead to heavy weeping rate. At the same time, excessive gas should be avoided to prevent excessive entrainment rate causing flooding（选定液体流量后，应避免气体量太小，以防止过大的泄漏率；同时应避免气体量太大，以防止过大的雾沫夹带率造成液泛）.

③ The entrainment rate and weeping rate is obtained by measuring the entrainment amount and weeping amount over time using stopwatch timing.

V. Requirements for experimental report

(1) Plot the relation graphs of the pressure drop and superficial gas velocity on the dry tray or under various liquid flows on the log-log coordinate paper.

(2) Plot the relation graph between weeping rate, entrainment rate with superficial gas velocity on the log-log coordinate paper.

(3) Draw the capacity performance chart of the tray, and outline the reasons.

(4) Describe several flow states on the tray as the gas velocity changes from small to large.

VI. Questions

① Try to analyze the influence of gas load on the relationship of pressure drop to superficial gas velocity (Δp-u), entrainment rate to superficial gas velocity (e-u), weeping rate to superficial gas velocity (θ-u).

② Try to qualitative analyze the factors influencing the liquid flooding.

③ How will the capacity performance chart change with tray spacing increasing?

④ What is the main factor determining the flow state on the tray?

Experiment 10　Absorption-desorption experiment
实验10　吸收-解吸实验

I. Experiment purpose

① Familiar with the structure and operation of the packed absorption tower.

② Observe the hydrodynamic state of packed tower, and measure the relationship between pressure drop and gas velocity.

③ Master the determination method of the overall mass-transfer coefficients $K_x a$ of liquid volume and the height of a transfer unit H_{OL}, and analyze the influencing factors.

II. Experimental principle

(1) Hydrodynamic characteristics of the packed tower（填料塔流体力学特性）

The pressure drop and gas velocity of the packed tower are important hydrodynamic parameters in the tower design and operation. The relation between the pressure drop caused by gas passing through the packing layer and superficial gas velocity（空塔气速）is shown in Fig. 2-74. When there is no liquid spray, the relationship between the pressure drop of the dry packing Δp and gas velocity u can be seen as a straight line with a slope of 1.8-2 (line a-a' in figure) in the double logarithmic coordinates. At low gas velocity（before point c）, when liquid is sprayed, there is no obvious effect on the thickness of liquid film covering on packing surface. The liquid holding capacity in the packing layer is irrelevant to the gas velocity of

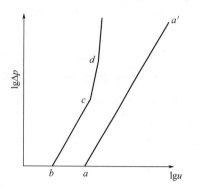

Fig. 2-74 Schematic diagram of packing pressure drop and air velocity of empty tower

图 2-74 填料层压降-空塔气速关系示意图

the empty tower, but only increases with the increase of spraying amount. The pressure drop is proportional to the gas velocity to the power of 1.8-2. Because the porosity of the packing layer reduces by liquid holding, the pressure drop is higher than that of the dry packing layer at the same gas velocity, indicating that the bc section in the figure is a constant holding zone（恒持液区）. As the gas velocity increases, the liquid film thickens, and the "blocking state"（or liquid loading phenomenon）["拦液状态"（或称载液现象）] with the increase of liquid holding capacity of the packing layer appears. The point c in the figure is the loading point or the liquid blocking point（载点或拦液点）. When the gas velocity is greater than the carrier gas velocity at the loading point, the liquid holding capacity in the packing layer increases with the increase of the gas velocity. The slope of the relation line between pressure drop and gas velocity increases, and the segment cd in the figure is the liquid-carrying section（载液区段）. The gas velocity continues to increase and reaches the point d（which is called the flooding point 泛点）, the gas velocity corresponding to this point is called the flooding gas velocity or the flooding point gas velocity. At this point, the drag force generated by the updraft on the liquid seriously impedes the downward flow of the liquid. The accumulated liquid fills the void of the packing layer resulting in a sharp rise in the pressure drop of the packing layer and a steepening of relation line between pressure drop and gas velocity. The line segment above point d in the figure is the flooding zone.

The actual gas velocity of packed tower is controlled to be close to the liquid flooding but does not occur. The mass transfer efficiency is the highest at this time. Generally, the operating gas velocity is 60%-80% of the flooding gas velocity（液泛区段）.

(2) Measurement of the overall mass-transfer coefficients $K_x a$ of liquid volume（液相

体积总吸收系数 $K_x a$ 的测定)

Desorption process is the reverse process of absorption, and the mass transfer directions are opposite. On the x-y diagram, the operating line of the absorption process is above the equilibrium line, while the operating line of the desorption process is below the equilibrium line. In this experiment, oxygen and water are in parallel flow contact in the absorption tower. Then oxygen-rich water formed by absorbing pure oxygen is sent to the top of the desorption tower to countercurrent contact air for desorption. In this experiment, the determination of the overall mass-transfer coefficients $K_x a$ of liquid volume under different liquid and gas quantities are required.

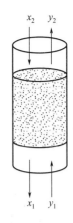

Fig. 2-75 Desorption test of oxygen-rich water

图 2-75 富氧水解吸实验装置

In this experiment, oxygen-rich water is desorbed, as shown in Fig. 2-75. Because the concentration of oxygen-enriched water is very low, it can be considered that the equilibrium relation of gas-liquid two-phase follows Henry's law, that is, both the equilibrium line and the operating line are a straight line. Therefore, the average driving force of mass transfer in packing layer can be calculated by logarithmic mean concentration difference. The corresponding mass transfer rate equation is as follows:

$$G_A = K_x a V_p \Delta x_m \tag{2-72}$$

That is:

$$K_x a = \frac{G_A}{V_p \Delta x_m} \tag{2-73}$$

Where:

$$\Delta x_m = \frac{(x_2 - x_2^*) - (x_1 - x_1^*)}{\ln \frac{x_2 - x_2^*}{x_1 - x_1^*}}$$

$$G_A = L(x_2 - x_1)$$

$$V_p = Z\Omega$$

The fundamental calculation equation of packing height (填料层高度) is:

$$Z = \frac{L}{K_x a \Omega} \int_{x_1}^{x_2} \frac{\mathrm{d}x}{x^* - x} = H_{OL} N_{OL} \tag{2-74}$$

That is

$$H_{OL} = \frac{Z}{N_{OL}} \tag{2-75}$$

Where,

$$N_{OL} = \int_{x_1}^{x_2} \frac{\mathrm{d}x}{x^* - x} = \frac{x_2 - x_1}{\Delta x_m}$$

$$H_{OL} = \frac{L}{K_x a \Omega}$$

G_A—desorption amount of oxygen per unit time, kmol/(m² · h);

$K_x a$—total mass transfer coefficient of liquid phase volume, kmol/(m³ · h);

V_p—packing volume, m³;

Δx_m—logarithmic mean concentration difference of liquid phase;

x_2—the mole fraction of inlet liquid phase (top of the tower);

x_2^*—the mole fraction of liquid phase in equilibrium with the outlet gas phase y_2 (top of the tower);

x_1—the mole fraction of outlet liquid phase (bottom of the tower);

x_1^*—the mole fraction of liquid phase in equilibrium with the inlet gas phase y_1 (bottom of the tower);

Z—packing height, m;

L—the flow rate of desorption liquid, kmol/(m² · h);

H_{OL}—the height of total transfer units based on the overall driving force for the liquid phase, m;

N_{OL}—the number of total transfer units based on the overall driving force for the liquid phase.

Oxygen is an indissolvable gas with very low solubility in water, so the mass transfer resistance is almost all concentrated in the liquid film, namely $K_x \approx k_x$. Since it is a liquid film control process, in order to increase the total mass transfer coefficient $K_x a$ of liquid volume, the degree of turbulence in the liquid phase should increase, that is, increase the spraying amount.

Both the equilibrium line and operation line are straight lines in the y-x diagram, so the unit of gas-liquid concentration in the calculation is the mole fraction (y, x) rather than the mole ratio (Y, X).

The total mass transfer coefficient of liquid volume $K_x a$ and the height of total mass transfer unit H_{OL} of liquid phase are as follows:

① Air flow rate during use（使用状态下的空气流量）

$$V_2 = V_1 \frac{p_1 T_2}{p_2 T_1} \tag{2-76}$$

Where, V_1—indicating value of air rotor flowmeter, m³/h;

T_1, p_1—temperature and pressure of air under calibration condition (20℃, 1.013×10⁵Pa);

T_2, p_2—the temperature and pressure of the air during use.

② Desorption of oxygen per unit time G_A（单位时间氧的解吸量G_A）

$$G_A = L(x_2 - x_1) \tag{2-77}$$

Where, L—flow rate of water, kmol/h;

x_1, x_2—the mole fraction of inlet or outlet liquid phase.

③ y_1 of the inlet gas phase concentration, and y_2 of the outlet gas phase concentration

$$y_1 = y_2 = 0.21$$

④ Logarithmic mean concentration difference Δx_m （对数平均浓度差 Δx_m）

$$\Delta x_m = \frac{(x_2 - x_2^*) - (x_1 - x_1^*)}{\ln \dfrac{x_2 - x_2^*}{x_1 - x_1^*}} \tag{2-78}$$

$$x_1^* = \frac{y_1}{m}$$

$$x_2^* = \frac{y_2}{m}$$

$$m = \frac{E}{p} \tag{2-79}$$

Where, m—phase equilibrium constant;

p—total pressure of the system, p = atmospheric pressure + 1/2 (packing pressure difference), kPa;

E—Henry's coefficient, kPa.

Henry's coefficient E at different temperatures of oxygen is obtained by the following equation:

$$E = (-8.5694 \times 10^{-5} t^2 + 0.07714 t + 2.56) \times 10^6 \tag{2-80}$$

⑤ $K_x a$ of total mass transfer coefficient of liquid volume （液体体积总传质系数 $K_x a$）

$$K_x a = \frac{G_A}{V_p \Delta x_m} \tag{2-81}$$

Where, $K_x a$—total mass transfer coefficient of liquid volume, $kmol/(m^3 \cdot h)$;

V_p—packing volume, m^3.

⑥ H_{OL} of height of total liquid mass transfer units （液相总传质单元高度 H_{OL}）

$$H_{OL} = \frac{L}{K_x a \Omega} \tag{2-82}$$

Where, H_{OL}—height of total liquid mass transfer unit, m;

L—flow rate of water, kmol/h;

Ω—cross-sectional area of packing tower, m^2.

III. Experimental device and process

(1) Schematic diagram of experimental equipment and flow chart

The diameter of desorption tower $\phi = 0.1$m, the diameter of absorption tower $\phi = 0.032$m, and the height of the packing layer is 0.8m. The packing characteristic parameters are shown in Appendix 6.

The experimental flow of oxygen absorption and desorption is shown in Fig. 2-76. Oxygen supplied by the oxygen cylinder enters the oxygen buffer tank through the pressure relief valve. The pressure is stabilized at 0.04-0.05MPa. For safety, a safety valve is installed on the buffer tank, which will open automatically when the pressure in the buffer tank reaches 0.08MPa. The flow rates of oxygen and water are regulated by the control valve and measured by the rotor flowmeter before entering the absorption tower. Oxygen is in parallel contact with water in the absorption tower to form heavy-oxygen-enriched water, which is sprayed from the top of the desorption tower through the pipe. The air is supplied by the fan and flows through the buffer tank into the bottom of the desorption tower. The air flow rate can be adjusted by a control valve and measured with an air rotor flowmeter. The air contacts with the heavy-oxygen-enriched water sprayed from the top of the desorption tower to desorb the oxidizing water. The exhaust gas after desorption is discharged from the top of the tower, and the oxygen-poor water is discharged from the bottom of the tower through the balance tank.

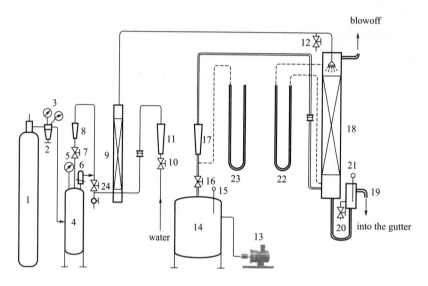

Fig. 2-76 Flow chart of oxygen absorption and desorption apparatus
图 2-76 氧气吸收与解吸实验装置流程图

1—Oxygen cylinder（氧气钢瓶）；2—Oxygen pressure relief valve（氧减压阀）；3,5—Oxygen pressure gauge（氧压力表）；4—Oxygen buffer tank（氧缓冲罐）；6—Safety valve（安全阀）；7—Oxygen flow control valve（氧气流量调节阀）；8—Oxygen rotor flowmeter（氧转子流量计）；9—Absorption tower（吸收塔）；10—Water flow control valve（水流量调节阀）；11—Water rotor flowmeter（水转子流量计）；12—Heavy-oxygen-enriched water sampling valve（富氧水取样阀）；13—Fan（风机）；14—Air buffer tank（空气缓冲罐）；15,21—Thermometer（温度计）；16—Air flow control valve（空气流量调节阀）；17—Air rotor flowmeter（空气转子流量计）；18—Desorption tower（解吸塔）；19—Liquid level balance tank（液位平衡罐）；20—Oxygen-poor water sampling valve（贫氧水取样阀）；22—Differential pressure gauge（压差计）；23—Gauge before flow meter（流量计前表压计）；24—Waterproof backfill valve（防水倒灌阀）

The gas flow rate is related to the gas state, so a gauge and thermometer are installed in front of each gas flow meter. In order to measure the pressure drop of packing layer, the desorption tower is equipped with a pressure differential meter.

An inlet oxygen-enriched water sampling valve is arranged at the inlet of the desorption tower to collect the inlet water sample, and the outlet water sample is sampled by the oxygen-poor water sampling valve on the bottom discharge balance tank. The liquid oxygen concentration of the water sample was measured by an oxygen meter. See Appendix 7 for its use.

（2）Experimental automatic control interface（Fig. 2-77）

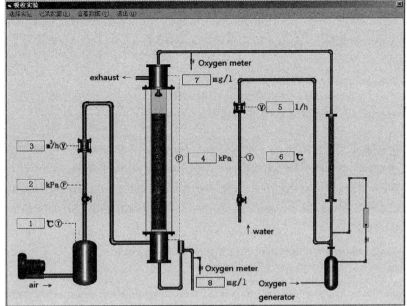

Fig. 2-77 Absorption-desorption experiment—Experiment automatic control interface
图 2-77 吸收-解吸实验——实验自控界面

IV. Experimental steps

（1）Experimental operations

① Determination of hydrodynamic properties of packed tower（填料塔流体力学性能测定）

a. Determine the pressure drop of dry packing（测定干填料压降）

ⅰ The packing in the tower must be dried in advance;

ⅱ Measure 6-8 sets of the pressure drop of the packed tower at different air flow rate.

b. Determine the pressure drop of wet packing（测定湿填料压降）

ⅰ Pre-flooding before the determination, so that the packing surface fully wetting.

ⅱ Under fixed spray amount, measure 8-10 sets of the pressure drop of the packed tower under different air flow.

ⅲ When the experiment approaches liquid flooding, the entering gas should not be too large, otherwise the flooding points in Fig. 2-74 is not easily to find（实验接近液泛时，进塔气体的增加量不要过大，否则不容易找到图 2-74 中的泛点）. Closely observe the gas-liquid contact status on the packing surface, and pay attention to the variation range of the pressure drop in the packing layer. Read the data after each parameter is stable. The pressure drop of packing layer is almost unchanged after the liquid flooding, but the gas velocity increases significantly. Slightly increase the gas amount, and take one or two more points. It should be noted that the gas velocity should not exceed the flooding point too much to avoid breaking through and rushing away the packing（注意不要使气速过分超过泛点，避免冲破和冲跑填料）.

② Determination of liquid volume total mass transfer coefficient $K_x a$（液相体积总传质系数 $K_x a$ 的测定）

a. Oxygen enters the buffer tank after decompression. The pressure in the tank is kept at 0.04-0.05MPa. The pressure should not be too high, and pay attention to the use of relief valve. To prevent water from pouring backward into the oxygen rotor flowmeter, the anti-pouring tank valve should be closed before opening the water valve, or first pass into oxygen and then water.

b. Selection of experimental operating conditions

$$\text{Water spray density(水喷淋密度)} = \frac{\text{Liquid flow rate}(m^3/h)}{\text{Cross sectional area of tower}(m^2)}$$

The spray density of water is 10-15$m^3/(m^2 \cdot h)$, the air velocity of the empty tower is 0.5-0.8m/s, and oxygen intake flow is 0.01-0.02m^3/h. The oxygen flow is properly adjusted to control the concentration of the absorbed oxygen-rich water no more than 19.9mg/L.

c. Determination of liquid oxygen concentration at the top and bottom of the tower: take out oxygen-rich water and oxygen-poor water from the top and bottom of the tower respectively, and analyze the oxygen content with an oxygen meter.

d. After the completion of the experiment, the oxygen cylinder valve should be closed first, and then close the oxygen relief valve and oxygen flow regulating valve. Check the main power supply, the main water valve and each pipeline valve.

(2) Operating notes

① During the hydrodynamic performance measurement of the packed column, it is not necessary to start the oxygen system, but only water and air.

② Note that the airflow-regulating valve should be slowly open and closed to avoid breaking the glass pipe（注意要缓慢开启和关闭空气流量的调节阀，以免撞破玻璃管）.

V. Requirements for experimental report

(1) Calculate and determine the relation of pressure drop per unit packing height of dry packing and wet packing with superficial gas velocity u. Plot correlation curve of $\Delta p/Z$-u in log-log coordinate system, find out loading point and flooding point.

(2) Calculate total mass transfer coefficient $K_x a$ and the liquid volume total mass transfer unit height H_{OL} under a certain spray amount and superficial gas velocity.

VI. Questions

① What are the structural features of packed tower?

② What is the purpose of measuring the relation curve of Δp-u? What is the relationship between the actual operating gas velocity and that at flood point?

③ In industry, why should absorption be carried out under low temperature and pressurized conditions, while desorption be carried out under high temperature and atmospheric pressure?

④ Why does the absorption and desorption of soluble gas belong to the gas film control process, while that of insoluble gas belong to the liquid film control process?

⑤ What are the characteristics of the packed tower structure?

Experiment 11 Drying experiment
实验 11 干燥实验

I. Experiment purpose

① Understand the structure characteristics of tunnel dryer and drying operation.

② Measure the drying curve, the drying rate curve and critical moisture content of the material under constant drying condition.

③ Determine the convective heat transfer coefficient of the material-air during the constant-rate drying step.

II. Experimental principle

The drying process is the operation to remove the relatively small amounts of water or other liquid from the solid material to reduce the content of residual liquid to an acceptably low value. In the drying process, heat transfer and mass transfer are conducted simultaneously. The mechanism is relatively complicated. By far, the drying rate is still determined by the experimental method.

Assume that the temperature, humidity, and velocity and direction of flow of the air across the drying surface are constant. This is called drying under constant drying conditions（恒定干燥条件）. The drying curve（干燥曲线）and drying rate curve（干燥速率曲线）are determined under constant drying conditions.

（1）The drying curve（干燥曲线）

The relation curve of free moisture content X versus time τ is the drying curve for constant drying conditions as shown in Fig. 2-78.

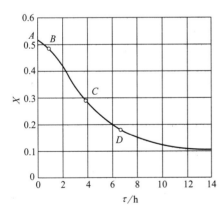

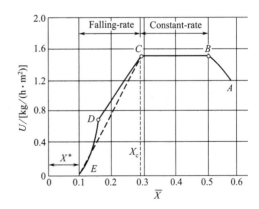

Fig. 2-78 Typical drying curve
图 2-78 典型的干燥曲线

Fig. 2-79 Typical rate-of-drying curve
图 2-79 典型的干燥速率曲线

（2）The rate-of-drying curve（干燥速率曲线）

The rate-of-drying curve for constant drying conditions is plot of data as drying rate versus material moisture content as shown in Fig. 2-79. At zero time, the initial free moisture content is shown at point A. Segment AB is the material heating period. This initial unsteady-state adjustment period is usually quite short, so it is generally incorporated into the BC segment or ignored in the analysis of times of drying. From point B to point C, the line is straight, and hence the slope and rate are constant during this period, which is called constant-rate drying period（恒速干燥阶段）. The drying rate of CDE period decreases with the decrease of material moisture content, so it is called the falling-rate period（降速干燥阶段）. The intersection point C between two drying periods is called the critical point（临界点）. The water content of the material corresponding to this point is called the critical moisture content（临界含水量）and expressed as X_c. The drying rate at this point is still equal to that at the constant-rate period, expressed as U_0. The material moisture content corresponding to point E is the equilibrium moisture content（平衡含水量）under the given air conditions, and the drying rate at this point is zero.

① Drying in the constant-rate period（恒速阶段干燥）

In the constant-rate drying period, the surface of the material is initially very wet and a continuous water film exists on the drying surface. So the drying rate is controlled by the surface moisture evaporation rate of the material, and hence this period is also called the surface evaporation control period（表面汽化控制阶段）. The heat transferred to the material by the drying medium is all used for evaporation of moisture. The surface temperature remains constant as well as the partial pressure of water vapor, so the drying rate is constant.

② Drying in the falling-rate period（降速阶段干燥）

When the material is dried to the critical moisture content, the drying process will enter the falling-rate period. The entire surface is no longer wetted, and the rate of moisture supplied to the surface is lower than the rate of water gasification on the surface of the material. Because the drying rate is controlled by the transfer rate of moisture within the material, this period is also called the internal migration control period（内部迁移控制阶段）. As the moisture content of the material decreases gradually, the moisture migration rate in the material also decreases gradually, so the drying rate constantly decreases.

③ Drying rate U（干燥速率,U）

Drying rate refers to the mass of water vaporized per unit dry area in unit time. The drying rate can be calculated as follows:

$$U = \frac{\Delta W}{A \Delta \tau} = \frac{\text{mass difference between two successive weighing of material(lost mass)}}{\text{Drying surface area} \times \text{time interval between two weighing}}$$

(2-83)

Where, U—drying rate, $kg/(m^2 \cdot h)$;

W—vaporized water content, kg;

A—dry surface area, m^2;

τ—drying time, h.

④ Moisture content of material, X

In terms of dry-based moisture content:

$$X = \frac{G_s - G_c}{G_c} = \frac{\text{moisture content of material}}{\text{moisture-free mass of material}}$$
$$= \frac{\text{weighed mass of material} - \text{moisture-free mass of material}}{\text{moisture-free mass of material}} \tag{2-84}$$

Where, X—dry-based moisture content of wet material;

G_s—mass of wet material, kg;

G_c—moisture-free mass of material, kg.

In the drying rate curve, the average moisture content（平均含水量）of material corresponding to the drying-rate U is:

$$\overline{X} = \frac{X_i + X_{i+1}}{2} \tag{2-85}$$

Where, $\overline{X}$—mean moisture content of wet material at a certain drying rate;

X_i, X_{i+1}—moisture content at the beginning and end of the time interval $\Delta\tau$.

(3) Determination of convective heat transfer coefficient between material surface and air in constant rate drying period（恒速干燥阶段时物料表面与空气之间对流传热系数的测定）

Under constant drying conditions, the heat transfer process between the material surface and the air can be expressed as follows:

$$U_c = \frac{dW}{A d\tau} = \frac{dQ}{r_{t_w} A d\tau} = \frac{\alpha(t - t_w)}{r_{t_w}} \tag{2-86}$$

or
$$\alpha = \frac{U_c r_{t_w}}{t - t_w} \tag{2-87}$$

Where, α—convective heat transfer coefficient between material surface and air in constant rate drying period, W/(m² · ℃);

U_c—drying rate in constant rate drying period, kg/(m² · h);

t_w—wet bulb temperature of the air in the dryer, ℃;

t—dry bulb temperature of the air in the dryer, ℃;

r_{t_w}—heat of vaporization of water at t_w, J/kg.

For a stationary material, if the direction of air flow is parallel to the material surface, the convective heat transfer coefficient is generally calculated by the following empirical formula in the applicable range of $\overline{L} = 2450 - 29300$ kg/(m² · h) and air temperature of 45-150℃.

$$\alpha = 0.0204(\overline{L})^{0.8} \tag{2-88}$$

Where, $\overline{L}$—mass velocity of wet air [kg/(m² · h)].

And
$$\overline{L} = \frac{V_t \rho_m}{A} \tag{2-89}$$

Where, V_t—volume flow of air in dryer, m³/h;

ρ_m—the density at average air temperature t_m in the dryer (i.e. the average temperature of controlled temperature at the inlet of the dryer and the dry bulb temperature), kg/m³;

A—flow area of tunnel dryer, m²; [$A = 0.12 \times 0.12 = 0.0144$ (m²)].

Volume flow of air in dryer V_t: From the flow formula of the throttle flowmeter and ideal gas law, it can be deduced:

$$V_t = V_o \times \frac{273 + t_m}{273 + t_o} \qquad (2\text{-}90)$$

Where, t_m—the average air temperature in the dryer, that is, the average temperature of controlled temperature at the inlet of the dryer and the dry bulb temperature, ℃;

t_o—temperature at the flowmeter, ℃;

V_o—the flow rate of air under normal pressure and t_o, m^3/h.

According to the flow standard format:

$$V_o = 2.12 R^{0.51} \qquad (2\text{-}91)$$

Where, R—the reading of U-tube gauge, mmH_2O ($1 mmH_2O = 9.80665 Pa$).

III. Experimental device and process

(1) Schematic diagram of experimental equipment and flow chart

The process of the experimental device is shown in Fig. 2-80. Main technical parameters of the equipment are as follows:

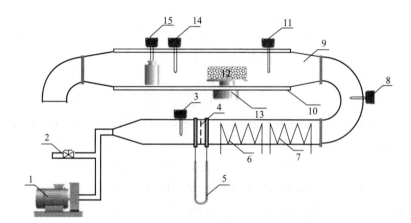

Fig. 2-80 Tunnel drying experiment—Device schematic diagram and flow chart

图 2-80 洞道干燥实验——实验装置示意图及流程

1—Vortex pump（旋涡泵）；2—Bypass flow control valve（旁路流量调节阀）；3—Platinum resistor of inlet flowmeter（流量计进口铂电阻）；4—Orifice meter（孔板流量计）；5—U-tube manometer（U 形压差计）；6—Constant-heating resistor wire（常热电阻丝）；7—Temperature-controlled resistor wire（控温电阻丝）；8—Inlet platinum resistor in dry section（干燥段进口铂电阻）；9—Tunnel（干燥风道）；10—Thermal insulation layer（保温层）；11—Platinum resistor for temperature control（控温铂电阻）；12—Material for drying（湿试样）；13—Weighing sensor（称重传感器）；14—Dry bulb platinum resistor（干球铂电阻）；15—Wet bulb platinum resistor（湿球铂电阻）

① Tunnel dryer: stainless steel, sandwich insulation, $120mm \times 120mm$;

② Vortex pump: maximum flow of 50 m^3/h;

③ Weighing sensor: maximum range of 500g, accuracy of 0.5 grade;

④ Orifice manometer: application range of 20-50m^3/h (U-tube differential pressure meter range of 600mmH_2O);

⑤ Temperature measurement: platinum resistor;

⑥ Heater: two sets of resistance wire, 2kW in each group, temperature control accuracy is $<\pm 1.0℃$;

⑦ Wet material: canvas.

(2) Experimental simulation interface (Fig. 2-81)

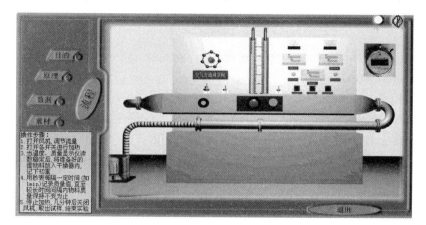

Fig. 2-81 Determination of tunnel drying curve—Experimental simulation interface
图 2-81 洞道干燥曲线测定实验——实验仿真界面

IV. Experimental steps

(1) Experimental operations

① Before the experiment

a. Read experiment instructor, and get familiar with the experimental equipment and process.

b. Check whether the wet bulb thermometer is installed correctly, and add water to the specified liquid level.

c. Dry the soaked sample in an electric drying oven, repeatedly weigh, until several weighing value are similar, which is regarded as the moisture-free mass of material（绝干物料质量）.

d. Adequately soak the sample, and note no water drops when it is placed into the dryer.

② Experimental steps

a. Start the fan and adjust the flow (e.g. adjusting the U-tube manometer reading to 200mmH_2O).

b. Turn on the heating resistance for preheating, generally adjust the temperature control current to 3-4A, the normal heat current to 1-2A, the temperature control set to 70℃.

c. After the zero point of the weighing sensor, the wet bulb temperature and the temperature control temperature are stable, record the flow rate, the temperature value of each tested point and the zero point value of the sensor. Gently put the prepared wet sample on the bracket of the sensor, and record the initial weight of the wet material.

d. Record the weight displayed by the weighing sensor every 1-2min. Continuous measurement is performed until the weight no longer shows obvious change.

e. Check the display value of dry bulb temperature and wet bulb temperature frequently. The change in the value indicates that the system is unstable or the wet bulb thermometer is short of water;

f. At the end of the experiment, stop heating, run the fan for a few minutes. When the temperature drops, stop the fan, remove the test sample and end the experiment.

(2) Operating notes

① The airflow should not be too large to prevent excessive noise.

② The heating current should not be too large to prevent damage to the resistance and the decline of temperature control accuracy.

③ The weighing sensor is easy to damage, so it must not be overweight. Never press by hand. (称重传感器极易损坏，使用时一定不能超重，严禁手压)

V. Record and organize experimental data

(1) Experimental data recording (Fig. 2-82)

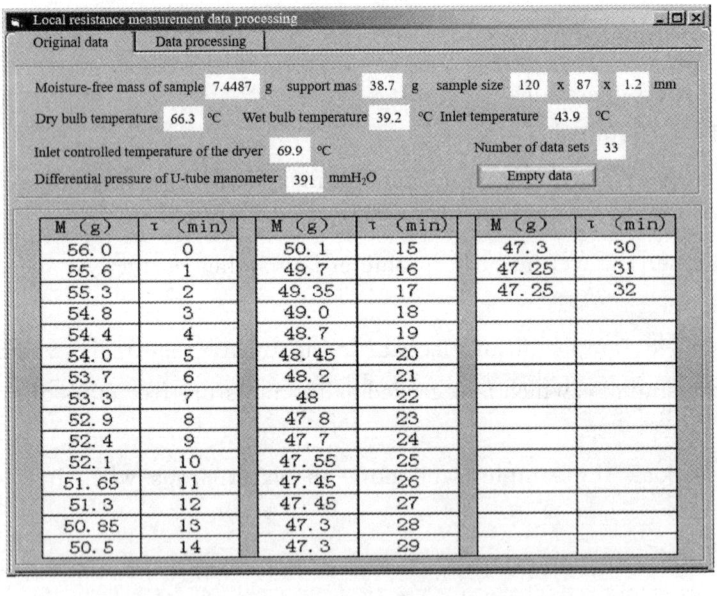
Local resistance measurement data processing

Original data | Data processing

Moisture-free mass of sample 7.4487 g support mas 38.7 g sample size 120 x 87 x 1.2 mm

Dry bulb temperature 66.3 ℃ Wet bulb temperature 39.2 ℃ Inlet temperature 43.9 ℃

Inlet controlled temperature of the dryer 69.9 ℃ Number of data sets 33

Differential pressure of U-tube manometer 391 mmH$_2$O Empty data

M (g)	τ (min)	M (g)	τ (min)	M (g)	τ (min)
56. 0	0	50. 1	15	47. 3	30
55. 6	1	49. 7	16	47. 25	31
55. 3	2	49. 35	17	47. 25	32
54. 8	3	49. 0	18		
54. 4	4	48. 7	19		
54. 0	5	48. 45	20		
53. 7	6	48. 2	21		
53. 3	7	48	22		
52. 9	8	47. 8	23		
52. 4	9	47. 7	24		
52. 1	10	47. 55	25		
51. 65	11	47. 45	26		
51. 3	12	47. 45	27		
50. 85	13	47. 3	28		
50. 5	14	47. 3	29		

Fig. 2-82 Determination of tunnel drying curve—Data recording

图 2-82 洞道干燥曲线测定实验——数据记录

(2) Experimental data collation (Fig. 2-83, Fig. 2-84)

VI. Requirements for experimental report

(1) Draw the drying curve and drying-rate curve on the ordinary coordinate paper, and determine the critical moisture content X_c and equilibrium moisture content X^*.

(2) Calculates the convective heat transfer coefficient between the material surface and

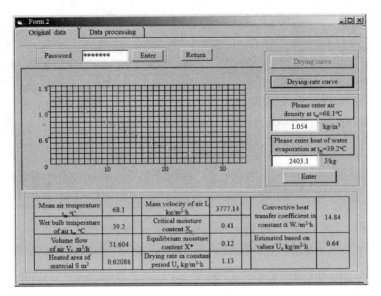

Fig. 2-83　Determination of tunnel drying curve—Data collection and plot（Drying curve）
图 2-83　洞道干燥曲线测定实验——数据整理、作图（干燥曲线图）

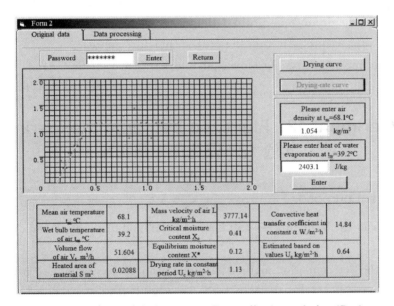

Fig. 2-84　Determination of tunnel drying curve—Data collection and plot（Drying-rate curve）
图 2-84　洞道干燥曲线测定实验——数据整理、作图（干燥速率曲线图）

the air during the constant-rate drying period according to the experimental results.

VII. Relevant materials（Fig. 2-85）

VIII. Questions

① In this experiment, how to keep the drying process under constant drying conditions?

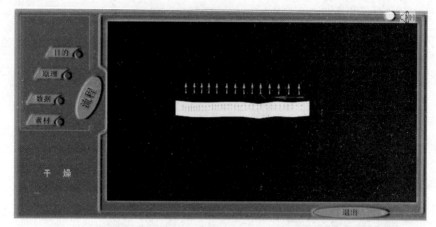

Fig. 2-85　Determination of tunnel drying curve—Demonstration of drying process
图 2-85　洞道干燥曲线测定实验——干燥演示

② After drying in the air flow at 70-80℃ for a long time, can absolutely dry materials be obtained? Explain the reason. How to obtain moisture-free materials?

③ Try to analyze the reasons why the drying rate remains constant during the constant rate drying period. What does the drying rate of the constant-rate drying period depend on?

④ What does the drying rate in the falling-rate drying period depend on?

⑤ What are the factors affecting the critical moisture content?

⑥ In the drying process, the blower should be turned on first, and then the electric heater. When the machine stops, it goes the reversed way. Try to explain the reason.

Experiment 12　Fluidized bed drying experiment
实验 12　流化床干燥实验

I. Experiment purpose

① Understand the basic process and operation method of fluidized bed dryer（流化床干燥器）.

② Master the determination method of fluidization curve of fluidized bed（流化床的流化曲线）and determine the relationship curve between bed pressure drop（床层压降）and gas velocity.

③ Determine the curve of relation between moisture content of materials and bed temperature with time.

④ Determine the drying rate curve of materials（干燥速率曲线）.

II. Experimental principle

The high-velocity flow makes the wet material in the boiling state, and the wet component in the material is vaporized and removed by controlling the temperature and speed of the air flow. This unit operation is fluidization drying（流化干燥）.

The characteristics of fluidized bed drying are as follows:

① High mass and heat transfer rate.

② The residence time of the material in the dryer can be adjusted freely, and the product with very low moisture content can be obtained.

③ Simple dryer structure, low cost, convenient operation and maintenance.

④ Suitable for handling particle materials with particle size of 30μm-6mm.

(1) Fluidization curve (流化曲线)

In the experiment, the bed pressure drop under different gas velocity can be measured, and the relationship curve between bed pressure drop and gas velocity can be obtained (see Fig. 2-86). At low air velocity, the operation is in the fixed bed stage (section AB). The bed is basically stationary, and the gas can only flow through the bed gap. The pressure drop is proportional to the flow velocity, with a slope of about 1 (in the double-log coordinate system). As the gas velocity gradually increases (into the BC segment), the bed begins to expand and the void ratio increases, the relationship between pressure drop and gas velocity will no longer be proportional.

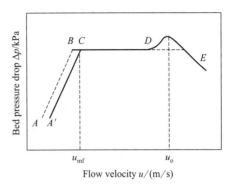

Fig. 2-86　Fluidization curve
图 2-86　流化曲线

The gas velocity continues to increase and enters the fluidization stage (CD section), where solid particles float along with the gas flow. With the increase of gas velocity, the bed height increases gradually, but the bed pressure drop basically remains unchanged, which equals to the bed net weight per unit area. When the gas velocity increases to a certain value (point D), the bed pressure drop will decrease, and the gas gradually takes the particles away. At this point, it enters the airflow transmission stage. The velocity at point D is referred to as the carrying velocity (带出速度, u_o).

Reducing the gas velocity in the fluidized state, the relationship between pressure drop and gas velocity returns to point C along the DC line in the Fig. 2-86. If the velocity of gas continues to decrease, the curve will not continue to change according to CBA, but rather along the CA' point, where C is known as the initial fluidization velocity (起始流化速度, u_{mf}).

An important characteristic of fluidized beds is that the gas velocity should be between the initial fluidization velocity and the carrying velocity, while the bed pressure drop remains constant in the production operation. Accordingly, the merits of bed fluidization can be judged by determining the bed pressure drop (可以通过测定床层压降来判断床层流化的优劣).

(2) Drying characteristic curve (干燥特性曲线)

Place the wet material under certain drying conditions to determine the relationship between the mass and temperature of the dry material over time, then the relation curve between moisture content (含水量, X)-time (τ) and material temperature (θ)-time (τ)

can be obtained (see Fig. 2-87). The slope of the curve between moisture content and time is drying rate (u). The drying rate curve (干燥速率曲线) is plotted by the drying rate (u) and the moisture content of materials ($\overline{x}$) (see Fig. 2-88).

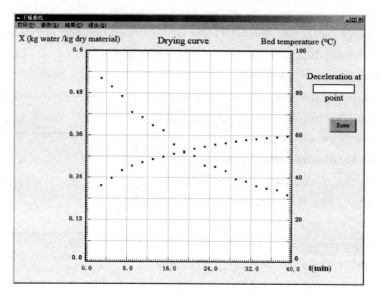

Fig. 2-87 Drying curve
图 2-87 干燥曲线

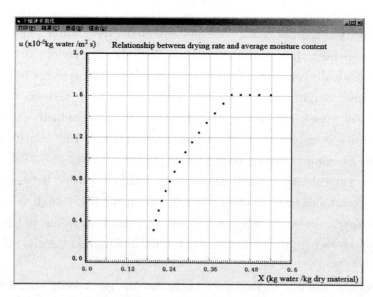

Fig. 2-88 Drying rate curve
图 2-88 干燥速率曲线

The drying rate is the water component vaporized per unit area per unit time, which is expressed by differential formula as:

$$u = \frac{dW}{A\, d\tau} \tag{2-92}$$

Where, u—drying rate, $kg/(m^2 \cdot s)$;

A — surface area of dry material, m²;

dτ — drying time, s;

dW — vaporized water component, kg.

To facilitate the calculation, the upper formula can converte into the following form:

$$u = \frac{\Delta X}{\alpha \Delta t} \tag{2-93}$$

Where, ΔX — variation of moisture content between two adjacent sampling points;

Δt — time interval between two adjacent sampling points, s;

α — specific surface area of mung bean = 1.23 m²/kg.

The moisture content of the material refers to the mass of water contained in the absolute dry material per unit mass, expressed in the following formula:

$$X_i = \frac{G_{si} - G_{ci}}{X_{ci}} \tag{2-94}$$

Where, G_{si} — the mass of the wet material taken out at moment i, kg;

G_{ci} — absolute dry mass of material taken out at moment i, kg;

X_{ci} — moisture content of material taken out at moment i.

Then the average moisture content is:

$$\overline{X} = \frac{X_i + X_{i+1}}{2} \tag{2-95}$$

Where, X_{i+1} — moisture content of material at moment $i+1$.

Drying characteristic curve can only be measured experimentally. Drying curve, drying rate curve and critical moisture content (临界含水量, X_c) are closely related to drying conditions (hot air flow rate, temperature, humidity, etc.), material characteristics, material size and contact state with hot air, etc. Therefore, for powdery materials, when each particle is dispersed and dried in hot air, not only the drying area is large, but also the critical moisture content is low, and it is easier to dry. If it is in the stacking state, the hot wind flows parallel through the accumulated material surface for drying, the critical moisture content increases and drying rate slows down. The same or similar experimental data with actual drying conditions and working conditions should be selected as the design reference. The experimental device is an intermittent fluidized bed dryer which can determine the required drying time to achieve a certain drying degree and provide the corresponding design parameters for the fluidization drying process operated continuously in industry.

During the fluidization drying, to ensure the material in the fluidization state, and to prevent the material from being carried out by the air flow, the operation gas velocity should be controlled between the fluidization gas velocity and the carrying velocity. At low air velocity, the operation is in the fixed bed stage. The bed is stationary, and the gas can only flow through the bed gap. The pressure drop is proportional to the flow velocity, with a slope of about 1 (in the double-log coordinate system). As the gas velocity gradually increases, the bed begins to expand and the void ratio increases. The material begins to boil, the bed pressure drop remains unchanged, which equals to the bed net weight per unit area. When the gas velocity increases to a certain value, the bed pressure drop will decrease. Gas gradually takes the particles away. At this point, it enters the airflow transmission stage (气流输送阶段).

Air flow is calculated by orifice flow meter with the formula:

$$V_s = 26.8 \Delta p^{0.54} \quad (2\text{-}96)$$

Where, Δp—pressure drop of orifice plate, kPa;

V_s—air flow, m³/h.

If the gas state of the orifice flowmeter varies from the calibration state (标定状态), the following formula is used to correct:

$$V_{s,2} = V_{s,1} \sqrt{\frac{p_1 T_2}{p_2 T_1}} \quad (2\text{-}97)$$

Where, $V_{s,1}$—indicated value of air rotameter, m³/h;

T_1, p_1—temperature and pressure of air in the calibrated state (293K, 101.3kPa);

T_2, p_2—temperature and pressure of air in use state.

III. Experimental device and process

(1) Schematic diagram of experimental equipment and flow chart

The flow of the fluidized bed drying experiment device is shown in Fig. 2-89.

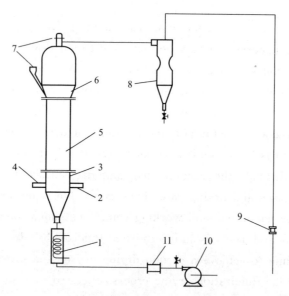

Fig. 2-89 The flow of the fluidized bed drying experiment device
图 2-89 流化干燥实验装置流程示意图

1—Air heater（空气加热器）；2—Discharge port（放净口）；3—Stainless steel cylinder（不锈钢筒体）；4—Sampling port（取样口）；5—Glass cylinder（玻璃筒体）；6—Gas-solid separation section（气固分离段）；7—Feeding port（加料口）；8—Cyclone separator（旋风分离器）；9—Orifice flowmeter（孔板流量计）$d_0 = 20$mm；10—Fan（风机）；11—Wet ball temperature water cylinder（湿球温度水缸）

All the equipment of this device is made of stainless steel, except the part of the bed cylinder body is made of high temperature hard glass. The cylinder body of the bed is composed of stainless steel section (inner diameter 100mm, high 100mm) and high temperature hard glass section (inner diameter 100mm, high 400mm), with gas-solid separation section (inner diameter 150mm, high 250mm) at the top. The stainless steel cylinder body is provided with material sampler, discharge port, thermometer connection, etc., respectively for sampling, purging and temperature measurement. The gas-solid separation section on the top of the bed body is equipped with feeding port and pressure tap for material feeding and pressure measurement respectively.

The air-heating device is composed of a heater and a controller. The heater is stainless steel coil heater; the outer wall is equipped with 1mm armored thermocouple. It functions with artificial intelligence instruments and solid state relays to realize the temperature control of air medium. At the same time, the instrument is controlled by computer. The air heater is equipped with dry bulb temperature and wet bulb temperature (干球、湿球温度) interfaces for air medium at the bottom.

The cyclone separator (旋风分离器) of this device can remove the dust of dry materials.

Each device is equipped with 8 instruments: heater temperature control, bed temperature, air temperature, air flow, air pressure, bed pressure drop, heating voltage and a multicircuit cycle detection meter (The red numbers in the simulation self-control interface are the dry bulb temperature and the green numbers are wet bulb temperature).

In this experiment, computer online data acquisition and control technology is introduced to speed up data recording and processing.

(2) Experimental automatic-control interface (实验自控界面) (Fig. 2-90)

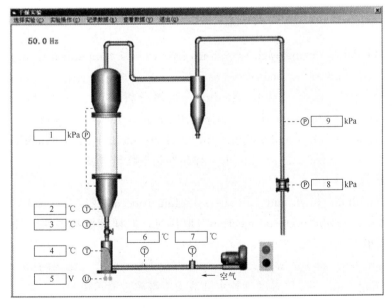

Fig. 2-90 Experiment of fluidized bed drying—Experimental automatic-control interface
图 2-90 流化干燥实验——实验自控界面

IV. Experimental steps

(1) Experimental operations

① Determination of fluidization curve（测定流化曲线）

a. Add solid material to the dryer or uses the remaining material from the drying experiment.

b. Open the fan, adjust the airflow and measure 8-10 groups of bed pressure drop under different airflow.

② Drying experiment（干燥实验）

a. Before the experiment, open the electronic balance and keep it in standby state.

b. Prepare a certain amount of dried materials（mung bean, for example）. Soak 0.5kg mung beans in hot water（60-70℃）for 20-30 minutes, take them out, blot the surface moisture with a dry towel, and set aside.

c. Fill the water tank of the wet bulb thermometer with water, but the liquid level shall not exceed the warning value.

d. Bed body preheating stage（床身预热阶段）. Start the fan and heater, control the air at a certain flow rate（the pressure difference of the orifice flowmeter is about 3kPa）, control the heater surface temperature（80-100℃）or air temperature（50-70℃）. Open the feed port, slowly pour the material to be dried, and close the inlet.

③ Determination of drying rate curve

a. Sampling: sample with sampling tube（push in or pull out）per 2-3 minutes. Place the sample in a small vessel, record the sample number and sampling time for analysis. 8-10 groups of data need to be completed. After the experiment, turn off the heater and fan power.

b. Record the data, during each sampling, record bed temperature, dry and wet bulb temperature of air, flow rate, bed pressure drop, etc.

c. Results analysis—oven analysis method（烘箱分析法）: Weigh the samples on the electronic balance for 9-10g each time and put them into the 120℃ oven for drying. After drying for 1h, it is taken out and weighed on an electronic balance, which can be regarded as the mass of moisture-free of the sample（样品的绝干物料质量）.

(2) Operating notes

① When sampling, push and pull the sampling tube quickly. The tube seat should be covered with cloth to avoid material ejection（取样时，取样管推拉要快，管槽口要用布覆盖，以免物料喷出）.

② The liquid level of the water refill cylinder of the wet bulb thermometer shall not exceed the warning value（湿球温度计补水缸液面不得超过警示值）.

③ Operate the electronic balance according to the operation instructions.

V. Requirements for experimental report

(1) Plot the Δp-u diagram of fluidized bed on log-log coordinate paper with airflow

velocity (m/s) as abscissa and bed pressure drop Δp as ordinate.

(2) Plot the relation curves of moisture content and drying time x-t and bed temperature and drying time t-τ on ordinary coordinate paper.

(3) Plot the drying rate curve and indicate the drying operating conditions.

VI. Questions

① Describe the advantages of fluidized bed drying operation.

② What are the characteristics of the fluidized bed pressure drop and gas velocity curves obtained in this experiment?

③ How to judge the abnormal phenomena of slugging and channeling flow in the operation of fluidized bed using bed pressure drop? How to avoid the abnormal phenomenon?

④ Dry and wet bulb thermometers are installed at the inlet of the heater in this experimental device. Assuming that the drying process is isoenthalpy process（等焓过程），please draw a schematic diagram of air state change.

Experiment 13　Extraction experiment
实验 13　萃取实验

I. Experiment purpose

① Understand the structure and characteristics of the rotary extraction tower（转盘萃取塔）.

② Master the operation of liquid-liquid extraction tower（液-液萃取塔）.

③ Master the measurement method of mass transfer unit height, and analyze impact of external energy on the mass transfer unit height and flux（传质单元高度和通量）in liquid-liquid extraction tower.

II. Experimental principle

Extraction（萃取）is a unit operation to separate the raw material liquid mixture using the different solubility of the components in the two liquid phases. A certain amount of extractant is added to the raw material liquid, and then stirred to fully mix with the extractant. The solute diffuses from the raw material liquid to the extraction material through the phase interface, so the extraction operation also belongs to the two-phase mass transfer process.

In this experiment, water was used as an extract（萃取剂）for extracting benzoic acid from kerosene, with a concentration of approximately 0.0020（kg benzoic acid/kg kerosene）. Water phase is the extraction phase（萃取相）[represented by E, also known as continuous phase or heavy phase（连续相或重相）in this experiment], and kerosene phase is raffinate phase（萃余相）[represented by R, also known as dispersed phase or light phase（分散相

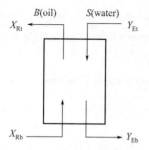

Fig. 2-91　Process diagram
图 2-91　流程示意图

S—Water flow rate（水流量）；
B—Oil flow rate（油流量）；
Y—Water concentration（水浓度）；
X—Oil concentration（油浓度）；
Subscript E—Extraction phase（萃取相）；
Subscript R—Raffinate（萃余相）；
Subscript t—Tower top（塔顶）；
Subscript b—Tower base（塔底）

或轻相）in this experiment〗. The light phase enters from the bottom of the tower, flows upward as the dispersed phase, and then flows out from the top of the tower after separation. The heavy phase flows downward from the top of the tower as a continuous phase to the bottom of the tower. The light and heavy phases flow in reverse（逆向流动）in the tower（see Fig. 2-91）. Benzoic acid partially transfers from the raffinate phase to the extractive phase during extraction. The concentrations of extraction phase and residual phase were determined by quantitative analysis. In view that water and kerosene are completely insoluble and the concentration of benzoic acid in both phases is very low, it can be considered that the volume flow rate of the two-phase liquid does not change during the extraction process.

III. Experimental device and process

The flow of extraction experiment device is shown in Fig. 2-92. In this experiment, benzoic acid was extracted from kerosene with water as extraction agent（本实验以水为萃取剂，从煤油中萃取苯甲酸）. Water is continuous phase（raffinate phase）, flowing from the water phase storage tank from the tower top through the water pump to the bottom of the liquid level adjustment tank. The kerosene phase is a dispersed phase（raffinate phase）, flows from the bottom through the oil pump to the light phase extraction tank. The mass of

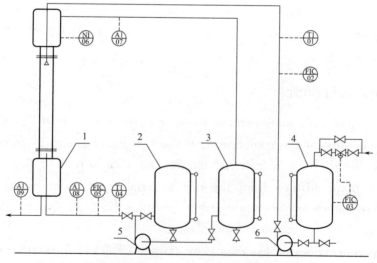

Fig. 2-92　Flow of rotary extraction tower
图 2-92　转盘萃取塔流程

1—Extraction tower（萃取塔）；2—Light phase liquid tank（轻相料液罐）；3—Light phase production tank（轻相采出罐）；4—Water phase storage tank（水相贮罐）；5—Light phase pump（轻相泵）；6—Water pump（水泵）

water phase and the oil phase is transferred by means of the rotation of the turntable in the tower.

Main technical parameters

① Tower diameter: 50mm; tower height: 750mm; effective height: 600mm; number of turntables: 16; turntable spacing: 35mm; turntable diameter: 34mm; fixed ring diameter: 36mm.

② Water pump and oil pump: Model—MG0XPS17.

③ Rotor flowmeter: Model—LZB-4; flow rate—16L/h.

④ Rotary governor: Type—6 Ik120RGK-CKF.

(1) Calculate the number of mass transfer units N_{OR} according to the raffinate（按萃余相计算传质单元数 N_{OR}）:

$$N_{OR} = \int_{x_2}^{x_1} \frac{dx}{x - x^*} \quad (2-98)$$

Where, N_{OR}—the total number of mass transfer units based on the raffinate phase;

x—concentration of solute in the raffinate phase, expressed by molar fraction;

x^*—concentration of solute in the raffinate phase in equilibrium with the corresponding extraction concentration, expressed as molar fraction;

x_1, x_2—represent the raffinate concentration of the two-phase inlet and outlet columns respectively.

The relationship of $\frac{1}{x-x^*}$-X can be obtained with the equilibrium curve and operating lines of benzoic acid in the two phases. Then N_{OR} can be obtained by diagrammatic integration or by Simpson numerical integration.

(2) Calculate the height of mass transfer units H_{OR} based on the raffinate（按萃余相计算传质单元高度 H_{OR}）:

$$H_{OR} = \frac{H}{N_{OR}} \quad (2-99)$$

Where, H_{OR}—the height of mass transfer units according to the raffinate, m;

H—effective contact height of extraction tower, m.

(3) Calculate the total mass transfer coefficient based on raffinate（按萃余相计算总传质系数）:

$$K_x a = \frac{L}{H_{OR} \Omega} \quad (2-100)$$

Where, $K_x a$—total mass transfer coefficient based on raffinate phase, kg/(m^3·h);

L—mass flow rate of raffinate, kg/L;

Ω—cross-sectional area of the tower, m^2.

If tower height H and mass transfer unit number N_{OR} are known, the value of H_{OR} can be obtained from the above equation. H_{OR} reflects the mass transfer performance of extraction device. The larger H_{OR}, the less the efficiency of the equipment. There are many factors affecting the mass transfer performance H_{OR} of the extraction device, mainly including device structure, two-phase material factors, operation factors and the form and size of the external energy. the total mass transfer coefficient $K_x a$ based on the raffinate can be calculated by H_{OR} according to equation (2-100).

(4) Experimental automatic-control interface（实验自控界面）(Fig. 2-93)

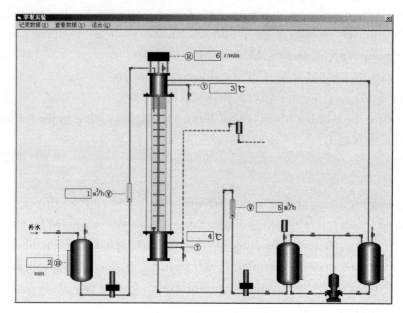

Fig. 2-93　Experiment of rotary disk extraction—Experimental automatic-control interface

图 2-93　转盘萃取实验——实验自控界面

IV. Experimental steps

(1) Experimental operations

① Inject an appropriate amount of water into the water storage tank, and add a prepared kerosene solution (e.g., 0.002kg benzoic acid per kg kerosene) into the oil material tank.

② Fully open water rotor flowmeter, and send the water as continuous phase into the tower. When the liquid level in the tower rises near to the midpoint of the heavy phase inlet and the light phase exit, adjust the water flow to a specified value (e.g., 4L/h) and slowly adjust the liquid level regulating tank to stabilize the liquid level.

③ Turn the turntable speed knob to zero, and then slowly adjust the speed to the set point.

④ Transfer the oil phase with the set flow rate (e.g., 6L/h) into the tower. Adjust the liquid level in time to keep stable near the midpoint of the heavy phase inlet and light phase outlet.

⑤ After half an hour of steady operation, collect 40mL of inlet and outlet oil-phase samples, and 50mL of outlet water-phase samples in conical bottles for concentration analysis.

⑥ Determine the concentration of each sample by quantitative analysis. Take 10mL kerosene solution and 25mL aqueous solution by pipette, titrate benzoic acid with 0.01mol/L NaOH standard solution using phenolphthalein as indicator. During titration, add several drops of dilute solution of non-ionic surfactant and violently shake until the end-point of titration.

⑦ After sampling, change the two-phase flow rate or rotary speed to determine the next experimental point.

(2) Operating notes

① During operation, absolutely avoid the two phase interface above the light phase outlet at the top of the tower (绝对避免塔顶的两相界面在轻相出口以上), it will cause the water phase to mix into the oil phase reservoir.

② Due to the large amount of dispersed phase and continuous phase retention at the top and bottom of the tower, after changing the operating conditions, the stabilization time must be long enough (about half an hour), otherwise the error will be large.

③ The actual volume flow of kerosene is not equal to the meter reading. When the actual flow value of kerosene is required, the reading of the flow meter must be corrected with the flow correction formula before being used.

V. Requirements for experimental report

(1) The number of mass transfer unit (N_{OR}), the height of mass transfer unit (H_{OR}) and the volume mass transfer coefficient ($K_x a$) were determined by changing the flow rate of oil phase with fixed rotation speed and water phase amount.

(2) The number of mass transfer unit (N_{OR}), the height of mass transfer unit (H_{OR}) and the volume mass transfer coefficient ($K_x a$) were determined by changing the rotation speed with fixed flow rate of oil phase and water phase amount.

VI. Questions

① What are the determinants of choosing to separate liquid mixtures by distillation or extraction separation?

② How does the operating temperature effect on the extraction fraction?

③ What is the effect of increasing solvent ratio on the extraction separation?

④ How to select the extractant?

⑤ Analyze the effects of rotating speed on mass transfer coefficient and extraction rate from experimental results.

⑥ What are commonly used extraction towers in industry? What are the main structural characteristics of the disk extraction tower?

Experiment 14　Pervaporation membrane experiment
实验 14　渗透蒸发膜实验

I. Experiment purpose

① Understand the separation principle of pervaporation (渗透蒸发).

② Master the operation of pervaporation separation of ethanol and water.

③ Study the main factors affecting pervaporation separation performance and their influencing laws.

II. Experimental principle

Membrane separation is an emerging and highly efficient separation technique. The separation process is carried out through the transfer medium, namely the membrane, under the action of a certain driving force (such as pressure difference, concentration difference, potential difference, etc.).

Pervaporation is a membrane osmosis process with phase transition. The pervaporation process is as follows: decompression at the downstream side of the membrane, the components, driven by the vapor pressure difference on both sides of the membrane, first selectively dissolve on the feed liquid surface of the membrane, then diffuse through the membrane, and finally gasify and desorption at the penetrating side surface of the membrane (渗透蒸发是在膜的下游侧减压,组分在膜两侧蒸汽压差的推动下,首先选择性地溶解在膜的料液表面,再扩散透过膜,最后在膜的透过侧表面气化、解吸). Pervaporation allows very low solute content to pass through the membrane to achieve separation from a large number of solvents (渗透蒸发可使含量极低的溶质透过膜,达到与大量溶剂分离的目的). Compared with distillation and other traditional separation methods, it has unique advantages in the separation of azeotrope, mixture with similar boiling point and isomer. There are also remarkable technological, equipment and economic advantages in the dehydration of organic solvents or mixed solvents containing a small amount of water, as well as in the treatment of wastewater containing small amounts of organic pollutants. Obviously, the separation of liquid mixture by pervaporation technology has the advantages of simple process, convenient operation, high efficiency, low energy consumption and no pollution.

After the liquid mixture raw material is heated to a certain temperature, it is fed into the film separator under atmospheric pressure. Low pressure of the downstream side of the membrane is maintained by vacuuming. The permeate component is driven through the membrane by the vapor partial pressure difference (or chemical potential gradient) on both sides of the membrane, and vaporizes on the downstream side of the membrane, where it is condensed into liquid and removed. The intercept (截留物) that cannot pass through the membrane can flow out of the separator from the upstream side of the membrane. The dissolution and diffusion of permeate components play an important role throughout mass transfer process. The evaporation transfer resistance at the penetrate side is relatively smaller and usually negligible, so the process is mainly controlled by the dissolution and diffusion steps.

The main indexes of pervaporation process are separation factor α (分离因子) and pervaporation flux J (渗透通量). The separation factor is defined as the composition ratio of the two components in the permeation liquid to the composition of the raw material liquid, which reflects the selective permeability of the membrane to the component.

The separation factors are calculated as follows:

$$\alpha = \frac{\gamma_A(1-C_A)}{C_A(1-\gamma_A)} \tag{2-101}$$

Where, α — separation factor;

γ_A — penetration concentration（透过液浓度）;

C_A — concentration of raw material liquid.

$$C_A = \frac{C_{A1}+C_{A2}}{2} \tag{2-102}$$

Where, C_{A1} — concentration of raw material solution before experiment;

C_{A2} — concentration of the raw material solution at the end of the experiment.

Permeable flux is defined as the permeable mass of components per unit time per unit membrane area, which reflects the rate of components permeable through the membrane. Permeable flux can be obtained as follows:

$$J = \frac{m}{S\Delta t} \tag{2-103}$$

Where, J — permeable flux, $kg/(m^2 \cdot min)$;

m — permeable liquid mass, kg;

S — membrane area, m^2;

Δt — operation time, min.

III. Experimental device and process

(1) Experimental apparatus

The vacuum on the permeating side of the equipment is formed by suction by vacuum pump. The minimum pressure can reach 2kPa. The operating temperature of the membrane chamber is in the range of room temperature to 90℃.

(2) The experimental process

The experimental apparatus and procedures are shown in Fig. 2-94. The device is mainly composed of raw material tank, feed pump, membrane assembly, sampling bottle, permeability collection device, buffer tank and vacuum pump.

IV. Experimental steps

(1) Experimental operations

① A certain concentration of raw material liquid is provided in the raw material tank (95% alcohol in this experiment) to above 2/3 height of the liquid level meter to avoid dry burning damage of electric heater. Install the film into the film chamber, and tighten the bolts. Open the feed liquid heater. Adjust the feed liquid temperature to the appropriate value. Open the feed pump, and start to cycle the feed liquid, so that the feed liquid temperature and concentration tend to be uniform.

② Determine the concentration of raw material solution (C_{A1}) using gas chromatograph.

③ After weighed by an electronic balance (m_1), the permeability collection pipe is loaded into a cold trap, and then installed on the pipeline. Open the vacuum pipeline and

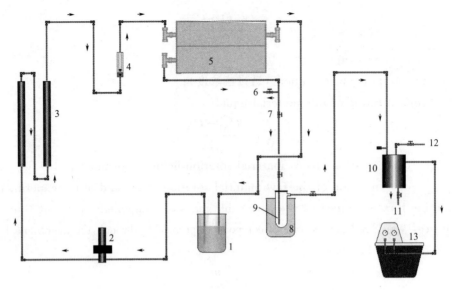

Fig. 2-94　Pervaporation experimental equipment and flow chart
图 2-94　渗透蒸发实验装置及流程示意图

1—Raw material tank（原料罐）；2—Feed pump（进料泵）；3—Heating rod（加热棒）；4—Rotor flow meter（转子流量计）；5—Membrane assembly（膜组件）；6—Vent valve（放空阀）；7—Flow regulating valve（流量调节阀）；8—Liquid nitrogen cooled trap（液氮冷阱）；9—Penetration fluid collection tube（渗透液收集管）；10—Buffer tank（缓冲罐）；11—Water discharge valve（放水阀）；12—Solenoid valve（电磁阀）；13—Vacuum pump（真空泵）

check the leakage.

④ When the feed liquid temperature is constant, open the vacuum pump and the vacuum pipe valve to observe the vacuum situation. After the pressure of the vacuum line reaches the predetermined value, the liquid nitrogen-cooling device is installed to start the pervaporation experiment. At the same time, record the data such as start time, material liquid temperature, permeate side pressure and feed liquid flow.

⑤ After the predetermined experimental time, turn off the vacuum pump, immediately remove the condense pipe and plug it (m_2). After the product melts, wipe the condensed water droplets on the outer wall of the condensate tube, and weigh (m_3). After the experiment, the concentration of the raw liquid (C_{A2}) and penetrating liquid (γ_A) were determined by gas chromatography.

⑥ Open the vent valve under the buffer tank in front of the vacuum pump, close the vacuum pump, close the feed pump, and end the experiment.

(2) Operating notes

① Understand the structure of the membrane assembly, and install the membrane correctly.

② Understand the process of vacuum system, and master the operation of vacuum pumping and draining. When stopping the vacuum operation, it is necessary to open the discharge valve to relieve pressure first, so as not to inhale the vacuum pump oil into the system.

③ When installing and removing the osmotic liquid collection pipe and applying liquid nitrogen cooling, special care should be taken to avoid being burned by liquid nitrogen.

④ Use chromatography and electronic balance correctly for component analysis and weighing operation.

V. Requirements for experimental report

(1) Measure the separation performance of pervaporation membrane, including its separation factor and flux.

(2) Observe the influence on the membrane separation performance of different feed temperature, feed composition, vacuum degree of downstream side of the membrane, etc., and analyze the results.

VI. Questions

① What are the general research contents and ideas of membrane process?

② What is the necessary cause for heating the feed liquid?

③ Compare the advantages and disadvantages of pervaporation and distillation.

Experiment 15 Ultrafiltration membrane separation experiment
实验 15　超滤膜分离实验

I. Experiment purpose

① Understand and be familiar with ultrafiltration membrane separation process.

② Determine the permeation flux and rejection rate of the experimental membrane.

II. Experimental principle

Ultrafiltration is a process of using small membrane pores to intercept macromolecules and separate liquid phase under the driving force of pressure difference. When the mixture containing high molecular solute (A) and low molecular solute (B) flows through the surface of the membrane under a certain pressure, the solvent and low molecular solute smaller than the membrane pore are collected as osmotic liquid (渗透液), while macromolecule solutes larger than membrane pores are recovered by membrane interface as concentrate (浓缩液). In general, the membrane separation process that can intercept molecular weight above 500 and below 10^6 is called ultrafiltration. The indexes to measure ultrafiltration membrane are rejection rate (截留率) f and permeation flux (渗透通量) $J[\text{L}/(\text{m}^2 \cdot \text{h} \cdot \text{kPa})]$.

The definition is as follows:

$$f = \frac{\text{Initial concentration of raw material liquid-concentration of permeable liquid}}{\text{Initial concentration of raw material liquid}} \times 100\% \tag{2-104}$$

$$J = \frac{\text{Volume of permeable liquid}}{\text{Experimental time} \times \text{membrane area} \times \text{pressure difference}} \tag{2-105}$$

III. Experimental device and process (See Fig. 2-94)

IV. Experimental steps

(1) Experimental operations

① Prepare an aqueous polyethylene glycol (PEG1000) with a concentration of 0.1% (mass ratio), and configure the solution of different concentrations by progressive dilution method. Taking distilled water as blank sample, the absorbance at different concentrations was measured by ultraviolet spectrophotometer, and the standard curve was drawn for analysis. (Structural principle and use of Type-752 spectrophotometer refer to Appendix 8)

② Install the membrane into the membrane assembly and tighten the bolt.

③ Measure the volume of the PEG feed liquid and add the feed liquid to the feed tank. After placing 5mL of feed fluid in a volumetric flask (50mL) with pipette and diluted to 50mL, measure the concentration of the feed fluid using a spectrophotometer.

④ Start the feed pump, ultrafilter the prepared PEG feed liquid under certain pressure and room temperature, measure the concentration and volume of PEG permeate. Stop the pump after the experiment.

(2) Operating notes

① The ultrafiltration membrane separation experiment is the same experimental device as the pervaporation membrane experiment, so it is not necessary to open the feed liquid heater for the ultrafiltration membrane separation experiment. （超滤膜分离实验与渗透蒸发膜实验为同一台实验装置，进行超滤膜分离实验时不需要打开料液加热器）

② After the experiment, clean the experimental device with tap water instead of raw liquid. Clean the residual PEG solution by running for a certain time at a large flow rate.

V. Requirements for experimental report

(1) Calculate the recovery rate y and the intercept rate f of the PEG:

$$y = \frac{\text{PEG content in concentrated solution}}{\text{PEG content in raw material solution}} \times 100\%$$

The intercept rate f can be calculated by referring to equation (2-104).

(2) Calculate the penetration flux J.

VI. Questions

① Try to discuss the mechanism of ultrafiltration membrane separation.

② What are the consequences of the excessive operating pressure or the flow in the experiment?

③ What is the effect of increasing the temperature of feed liquid for ultrafiltration?

Chapter 3　Methods of Processing Experimental Data

第 3 章　实验数据的处理方法

To compare and analyze with the theoretical values, the results after the calculation of experimental data are usually summarized into empirical formulas or in the form of charts. Therefore, the experimental data must be properly processed and analyzed to ensure that it is the right conclusion to stand up to testing. The basic knowledge of this area is described below.

3.1　Significant digits and operation rules （有效数字与运算规律）

(1) Significant digits （有效数字）

In measurement and experiment, we often encounter two kinds of numbers. One is unitless numbers, such as π, whose significant digits can be more or less according to our requirement. Another type is to represent the measurement results in unit, such as temperature, pressure, flow rate, etc. The last digit of such numbers is often estimated by the accuracy of the instrument. For example, a thermometer with an accuracy of 1/10℃ reads 21.75℃, the last digit is estimated. So recording or measuring data usually reserves a significant digit after the minimum scale of the instrument.

In science and engineering, in order to clearly represent the accuracy of the numerical value and to facilitate the operation, a decimal point is added after the first significant number while the order of magnitude of the value is represented by the power of 10. This method of counting by power of 10 is called scientific notation （用 10 的幂来记数的方法称为科学记数法）. For example, 185.2mmHg can be recorded as 1.852×10^2 mmHg.

(2) The operation rule of significant numbers （有效数字的运算规律）

① In addition and subtraction operations, the number of decimal places reserved for each number shall be the same as the number with the least number of decimal points （在加减运算中, 各数所保留的小数点后的位数应与其中小数点的位数最少的相同）. For example, $12.56 + 0.082 + 1.832 = 14.47$.

② In multiplication and division operation, the number of digits retained by each number is subject to the fewest significant numbers（在乘除运算中，各数所保留的位数以有效数字最少的为准）. For example，$0.0135 \times 17.5 \times 2.46 = 0.581$.

③ The result of power and square root operation retains one more significant digit than the original data（乘方及开方运算的结果比原数据多保留一位有效数字）. For example，$12^2 = 144$，$\sqrt{5.6} = 2.37$.

④ For logarithmic operation，the significant numbers are the same before and after the logarithm operation（对数运算，取对数前后的有效数字相等）. For example，$\lg 2.584 = 0.4123$，$\lg 2.5847 = 0.41241$.

3.2 Error analysis of experimental data（实验数据的误差分析）

The difference between the measured experimental value and the true value is called the error of the measured value. The estimation and analysis of measurement error are of great significance to the accuracy of experimental results.

(1) True and mean（真值与平均值）

Any measured physical quantity always has a certain objective true value, that is, the true value. Due to the errors caused by measuring instrument and method, etc., the true values are generally not directly measurable. In the case of infinite measurements in the experiment, according to the error distribution law, the probability of positive and negative errors is equal. By adding and averaging each measurement value, the approximate true value may be obtained without systematic error. Therefore, experimental science defines the truth value as: The average value of an infinite number of measurements is called a true value（实验科学将真值定义为：无限多次的测量平均值称为真值）. However, the actual number of measurements is limited，so the average value obtained by the finite number of measurements can only be the approximate true value, called the best value（有限测量次数求出的平均值，只能是近似真值，称最佳值）. In experimental measurement，the values measured by high-precision standard instruments are used instead of true values. Commonly used averages are in the following forms:

① Arithmetic mean（算术平均值）

$$x_m = \frac{x_1 + x_2 + \cdots + x_n}{n} = \frac{1}{n}\sum_{i=1}^{n} x_i \tag{3-1}$$

② Geometric mean（几何平均值）

$$x_c = \sqrt[n]{x_1 x_2 \cdots x_n} \tag{3-2}$$

③ Mean root mean square（均方根平均值）

$$x_s = \sqrt{\frac{x_1^2 x_2^2 \cdots x_n^2}{n}} = \sqrt{\frac{\sum_{i=1}^{n} x_i^2}{n}} \tag{3-3}$$

④ Logarithmic mean（对数平均值）

$$x_I = \frac{x_1 - x_2}{\ln \frac{x_1}{x_2}} \tag{3-4}$$

In the above calculations, x_1, x_2, ⋯, x_n = each measurement value; n = the number of measurements.

(2) Error representation（误差的表示方法）

① Absolute error（绝对误差）Δx

The difference between a measured value and the true value is called the absolute error. In the actual measurement, the true value is often replaced by the best value. The expression is as:

$$\Delta x = x_i - x \approx x_i - x_m \tag{3-5}$$

Where, Δx — absolute error;

x_i — the value of the i-th measurement;

x — the true value;

x_m — the average value.

② Relative error（相对误差）

The ratio of absolute error to true value is called relative error.

$$\delta = \frac{\Delta x}{x} \tag{3-6}$$

③ Fiducial error（引用误差）

The ratio of the absolute error of measurement to the full-scale value of the instrument is called fiducial error.

$$\text{fiducial error} = \frac{\text{maximum indicating error}}{\text{full scale value}} \tag{3-7}$$

Fiducial error is often used to indicate the accuracy of the instrument, which can be divided into several grades. When the percent sign of fiducial error is removed, the remaining value is called the accuracy grade of the instrument. The accuracy grade of measuring instruments is stipulated by the state uniformly. The accuracy grades of electrical instruments is divided into seven grades: 0.1, 0.2, 0.5, 1.0, 1.5, 2.5 and 5.0 respectively.

For example, the accuracy of a pressure gauge is 1.5, indicating that the maximum error of the instrument is 1.5% of the maximum range of the equivalent grade. If the maximum range is 0.4 MPa, the maximum error bit of the pressure gauge:

$$0.4 \times 1.5\% \text{ MPa} = 0.006 \text{ MPa} = 6 \times 10^3 \text{ Pa} \tag{3-8}$$

(3) Character and classification of errors（误差的性质及其分类）

① System error（系统误差）

When the same quantity is measured for many times under certain conditions, the value of the error always remains unchanged, or the error that changes according to a certain law is called the systematic error. For example, when measuring instruments with inaccurate calibration or zero-point uncalibrated instrument are used; Changes in experimental state and environment, such as external temperature, pressure, humidity; And the experiment operator's habit, bias and other factors will cause the system error. This kind of error is often in the measurement of the same physical quantity, its size and symbol are basically unchanged or follow a certain rule. It can be eliminated by accurate correction（这类误差往往在同一物理量的测定中其大小和符号基本不变或有一定的规律，经过精确的校正可以消除）.

② Random error（accidental error）［随机误差（偶然误差）］

When measuring the same physical quantity under the same conditions, the absolute error value fluctuates up and down, and the sign is sometimes positive, sometimes negative. There's no rule, and it's unpredictable, but the error is completely subject to statistical laws. For the same physical quantity, with the increase of the number of measurements, the arithmetic mean of the random error approaches to zero, so the arithmetic mean of the multiple measurements will be close to the true value.

③ Gross error（过失误差）

Gross error refers to the errors caused by operational errors or human errors. This kind of error is often shown as a big difference from the normal value, which should be eliminated in data collation（这类误差往往表现为与正常值相差很大，在数据整理时应予以剔除）.

(4) Accuracy of experimental data（实验数据的精确度）

The concept of accuracy and error is complementary. High accuracy, small error; the greater the error, the less the accuracy. It reflects the degree of systematic error and random error comprehensively. The repeatability of the data obtained in the measurement is called precision, which reflects the magnitude of random error. Taking target practice as an example, Fig. 3-1(a) shows dense impact points, but far from the bull's eye (true value), which indicate a high precision with small random error, but large systematic error. Fig. 3-1(b) shows low precision but high accuracy, that is, large random error but small systematic error. In Fig. 3-1(c), both the systematic error and random error are small as well as high accuracy.

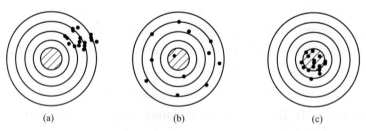

Fig. 3-1　Schematic diagram of precision and accuracy

图 3-1　精密度与精确度示意图

3.3　Experimental data processing
　　（实验数据处理）

The processing of experimental data is to express a series of experimental data in the most appropriate way after calculation and sorting. In chemical principle experiment, three forms are commonly used: tabular method, graphical method and equation representation.

(1) Tabular method (列表法)

The tabular method is to list the experimental data in a table according to the corresponding relationship between independent variables and dependent variables. With the advantages of easy tabulation, simplicity, compactness and easy comparison of data, tabulation method is the basis of plotting curves or sorting them into formulas. Experimental data are generally classified into experimental data table (original data table) and experimental data collation table.

The experimental data record table is designed according to the data to be tested. For example, the form of experimental data table of the straight-pipe resistance measurement experiment is as shown in Table 3-1.

Table 3-1　Experimental data on fluid resistance measurement through a straight pipe

表 3-1　流体直管阻力测定实验数据记录表

Pipe material _____; Inner diameter of pipe _____; Density of mercury _____;
Water temperature _____; Density of water _____; Viscosity of water _____

Serial number	The reading of the U-gauge			Indicated value of flow integrator
	Left side/mm	Right side/mm	The reading difference/mm	
1				
2				
3				
⋮				

The experimental data collation table is a tabular form indirectly derived from the calculation and sorting of experimental data, which expresses the relationship between the main variables and the experimental conclusions, as shown in Table 3-2.

Table 3-2　Experimental collation data on fluid resistance measurement through a straight pipe

表 3-2　流体直管阻力测定实验数据整理表

Serial number	Flow rate $V/(m^3/s)$	Velocity $u/(m/s)$	Reynolds number $Re \times 10^4$	The different reading of the U-gauge R/mmHg	Resistance loss $h_f/(J/kg)$	Friction coefficient λ
1						
2						
3						
⋮						

The following problems should be noted when drawing up the table according to the experimental content:

① Form design should be concise, clear, easy to read and use. The recorded and calculated items should meet the experimental requirements.

② Variable names, symbols and units should be titled at the header of each column, which should be designed clearly and reasonably.

③ The data in the table must reflect the accuracy of the instrument, and attention should be paid to the number of significant digits.

④ The number with large or small orders of magnitude can be recorded by the scientific notation. For example, $Re = 25000$ may be referred to as $Re = 2.5 \times 10^4$ in the scientific notation, which is recorded as $Re \times 10^4$ in the title bar and 2.5 in the data table.

⑤ Constant induction (that is, transforming factor) method should be used as much as possible in data collation to simplify calculation and reduce errors. For example, the Reynolds number at different velocity in a fixed pipeline is calculated by $Re = du\rho/\mu$. Since d/μ and ρ are definite values, it can be reduced to $Re = Au$, where constant $A = d\rho/\mu$ is the transforming factor. Multiplying A by different velocity u yields a series of corresponding Re, which reduces double computation.

⑥ Under the data collation table, a set of data is required as a calculation example to illustrate the relationship between the items for easy reading or checking.

(2) Graphic method (图示法)

The tabulation methods described above are often difficult to visualize the regularity of the data. In order to facilitate the comparison and show the regularity or changing trend of the results, the experimental results are often graphically represented. Some basic principles of correct mapping in chemical experiments are as follows:

① Selection of coordinates The coordinates commonly used in chemical industry include ordinary rectangular coordinates, logarithmic coordinates and semi-logarithmic coordinates (普通直角坐标、双对数坐标和半对数坐标). The appropriate coordinate paper is selected according to the functional relationship between variables.

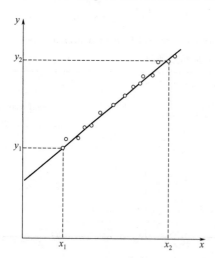

Fig. 3-2 Ordinary rectangular coordinate

图 3-2 普通直角坐标纸

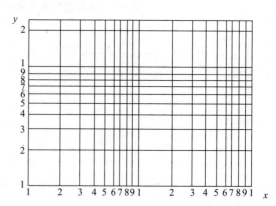

Fig. 3-3 Logarithmic coordinate

图 3-3 双对数坐标纸

a. As to linear function, $y = a + bx$ (as shown in Fig. 3-2), ordinary rectangular coordinate paper is used. Set up a set of experimental data whose variables fit the linear relationship described above. The intercept a and slope b in the equation are shown in the following graphic solution of monadic equation.

b. Logarithmic coordinates. The horizontal and vertical coordinates of the double logarithmic coordinate paper are on logarithmic scale, as shown in Fig. 3-3.

② Characteristics of logarithmic coordinates (对数坐标的特点) The distance between a point and the origin is the logarithmic value of the quantity represented at that point, but the quantity marked at the point is its own value (对数坐标的特点是：某点与原点的距离为该点表示量的对数值，但是该点标出的量是其本身的数值). For example: the distance from the point marked as 5 to the origin on the logarithmic coordinate is $\lg 5 = 0.7$, as shown in Fig. 3-4.

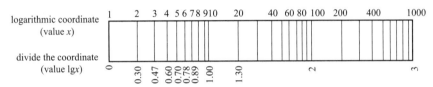

Fig. 3-4 Characteristics of logarithmic coordinates

图 3-4 对数坐标的特点

The upper line in Fig. 3-4 is the logarithmic scale of the x, while the lower line is the linear (uniform) scale of the $\lg x$. The actual distance between 1, 10, 100, 1000 on logarithmic coordinates is the same, because the corresponding logarithmic value of the above numbers are 0, 1, 2, 3, where the distance between the adjacent points is the same in linear (uniform) coordinates.

The distance on a logarithmic scale (measured with a uniformly scaled ruler) is expressed as the logarithmic difference of the values [在对数坐标上的距离（用均匀刻度的尺来量）表示为数值之对数差], i.e., $\lg x_1 - \lg x_2$:

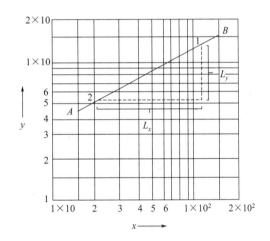

Fig. 3-5 Double logarithmic coordinates

图 3-5 双对数坐标纸上直线斜率和截距的求法

$$\lg x_1 - \lg x_2 = \lg \frac{x_1}{x_2} = \lg\left(1 - \frac{x_2 - x_1}{x_2}\right) \qquad (3\text{-}9)$$

Therefore, on the logarithmic coordinate paper, the distance between any two test points is the same as the straight-line distance on the drawing (refers to the uniform dividing scale), then the relative error between each point and the graph line is the same.

On the logarithmic graph paper, the slope of the straight line is:

$$\mathrm{tg}n = \frac{\lg y_2 - \lg y_1}{\lg x_2 - \lg x_1} \tag{3-10}$$

Since $\Delta \lg y$ and $\Delta \lg x$ are the distance L_x and L_y on the x and y axis respectively, L_x and L_y of the horizontal and vertical distance between 1 and 2 points on the straight line can be measured by ruler as shown in Fig. 3-5, then the slope is

$$n = \frac{\text{Measured } L_y \text{ of the vertical distance between points 1 and 2}}{\text{Measured } L_x \text{ of the horizontal distance between points 1 and 2}} \tag{3-11}$$

③ Basic principle of using double logarithmic coordinates（选用双对数坐标系的基本原则）

a. Applicable to power function $y = ax^n$, so that the nonlinear relationship converts into a linear relationship（适用于幂函数 $y = ax^n$，使非线性关系变换成线性关系）.

The power function is a curve plotted on the ordinary rectangular coordinates, which can be linearized by using double logarithmic coordinates. If taking logarithms on both sides of the above power function equation, then

$$\lg y = n \lg x + \lg a \tag{3-12}$$

Let: $\lg y = Y$, $\lg x = X$, and $\lg a = B$

The above equation converts into a linear equation $Y = nX + B$.

b. Applicable to the case where both the function y and independent variable x changed by several orders of magnitude（适用于所研究的函数 y 和自变量 x 在数值上均变化了几个数量级）.

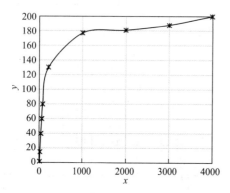

Fig. 3-6 Rectangular coordinates

图 3-6 用直角坐标纸做的图

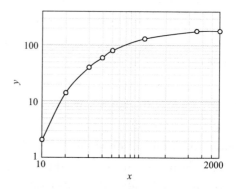

Fig. 3-7 Logarithmic coordinates

图 3-7 用双对数坐标纸做的图

For example, x and y are given as:

$x = 10$, 20, 40, 60, 80, 210, 1000, 2000, 3000, 4000

$y = 2$, 14, 40, 60, 80, 130, 177, 181, 188, 200

In rectangular coordinates, it is almost impossible to plot the points at the starting range of the curve when x is equal to 10, 20, 40, 60 or 80 (see Fig. 3-6), but a clearer curve can be obtained by using logarithmic coordinates (see Fig. 3-7).

④ The semi-logarithmic coordinates（半对数坐标系） One axis of the semi-logarithmic coordinate system is the ordinary coordinate axis with uniform indexing, and the other axis is the logarithmic coordinate axis（半对数坐标纸的一个轴是分度均匀的普通坐标轴，另一个轴是分度不均匀的对数坐标轴），as shown in Fig. 3-8.

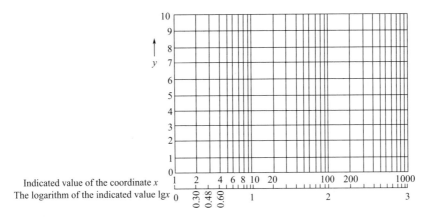

Fig. 3-8 Semi-logarithmic coordinates

图 3-8 半对数坐标系

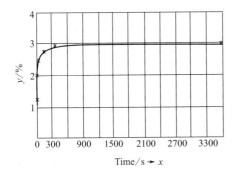

Fig. 3-9 Rectangular coordinates

图 3-9 用直角坐标纸作图

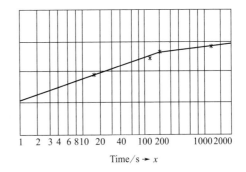

Fig. 3-10 Semi-logarithmic coordinates

图 3-10 用半对数坐标纸作图

The semi-log coordinates may be selected in the following cases：

a. One of the variables varies by several orders of magnitude in the studied scope.

b. In the initial stage where the independent variable gradually increases from zero, a little change of the independent variable causes a great change in the dependent variable. At this point, the use of semi-logarithmic coordinates can make the maximum range of curve elongation, so that the outline of the graph is clearer. For example, the diagram made of rectangular coordinate paper is shown in Fig. 3-9, while the semi-logarithmic coordinate paper is used in Fig. 3-10.

c. Applicable to exponential function $y = a\mathrm{e}^{bx}$, so that it is transformed into a linear function relationship. By taking the natural logarithm on both sides of the above equation, then $\ln y = \ln a + bx$, which is a linear relation between $\ln y$ and x.

⑤ The scale division of coordinates（坐标的分度） The scale division of coordinates refers to the value represented by each coordinate axis, that is, to choose the appropriate scale（坐标的分度是指每条坐标轴所代表数值的大小，即选择适当的比例尺）.

In the case where the error of x and y are given as Δx and Δy, to obtain an ideal figure, the scale should be obtained in such a way that the side length of the experimental "point" are $2\Delta x$ and $2\Delta y$, and make $2\Delta x = 2\Delta y = 1-2\text{mm}$, then

The scale of the x axis M_x is

$$M_x = \frac{2}{2\Delta x} = \frac{1}{\Delta x} \tag{3-13}$$

The scale of the y axis M_y is

$$M_y = \frac{2}{2\Delta y} = \frac{1}{\Delta y} \tag{3-14}$$

For example: the temperature error is known as $\Delta T = 0.05℃$, then

$$M_T = \frac{1}{0.05} = 20 \tag{3-15}$$

The temperature coordinate grading is 20mm long. If it feels too big, $2\Delta x = 2\Delta y = 1\text{mm}$ is advisable. At this point, the coordinate length of 1℃ is 10mm.

⑥ The use of coordinate paper and the plotting of experimental data（坐标纸的使用及实验数据的标绘）

a. According to the use habit, take the horizontal axis as the independent variable and the vertical axis as the dependent variable（取横轴为自变量，纵轴为因变量）. Then indicate the names, symbols and units represented by each axis.

b. Scale division of the coordinate axis according to the size of the plotting data（对坐标轴进行分度）. The so-called coordinate axis indexing is to select size of the numerical value represented by each scale on the coordinate. The minimum scale of the coordinate axis indicates the significant digits of the experimental data. The easy-to-read number is added to the scale line.

c. The selection of coordinate origin. In general, the origin of ordinary rectangular coordinate does not necessarily start from zero（普通直角坐标原点不一定从零开始）. Depending on the range of plotting data, the minimum data can be selected and move the origin to the appropriate location. For logarithmic coordinates, the axis scale is divided by the logarithmic value of $1,2,\cdots,10$, and each scale is the true number（对于对数坐标，坐标轴刻度是按 1、2、……、10 的对数值大小划分的，每刻度为真数值）. When using coordinates to represent different data, the scale division should follow the law of logarithmic coordinates. Each value can only be multiplied by $10n$ times (n is positive or negative integers), and cannot be arbitrarily divided. Therefore, the origin of the coordinate axis can only take the value specified on the logarithmic axis as the origin, and cannot be determined arbitrarily（坐标轴的原点只能取对数坐标轴上规定的值作原点，而不能任意确定）.

d. The plotted figure should fill up the whole coordinate paper and be symmetrical in the middle to avoid the drawing on one side.

e. Plotting data and curves: Plotting experimental results point by point on the coordinate paper according to the relationship between independent variables and dependent variables. If several groups of data are plotted on the same coordinate paper, the experimental points should be distinguished by different symbols (such as ●, ×, ▲, ○, ◆, etc.). A smooth curve is drawn according to the distribution of the experimental points. The curve should pass through the dense area of the experimental points so that the experimental points are as close as possible to the curve, and evenly distributed on both sides of the curve. Individual points that deviate far from the curve should be eliminated.

3.4 Representation of experimental data with mathematical equation

In the process of chemical experiment data processing, in addition to using tables and graphs to describe the relationship between variables, it is often necessary to express the experimental data or calculation results in the form of mathematical equations or empirical formulas.

In chemical engineering, empirical formulas are usually expressed as dimensionless number groups or quasi-number relations. The determination of constants and undetermined coefficients in the formula is the key to the equation representation of experimental data.

There are many ways to find the constants and undetermined coefficients in the empirical formula or the criterion relations. The most commonly used graphic method, point selection method, average method and least square method are introduced below.

(1) Graphic method (图解法)

The graphical method is limited to solving the constants of functional expression with linear relations or non-linear relations which are converted into linear relations (图解法仅限于具有线性关系或非线性关系式通过转换成线性关系的函数式常数的求解). The first step is to select the coordinate system and plot the experimental data into a straight line on the graph. By solving the slope and intercept of the line, the constants of the linear equation can be determined.

① Graphical representation of a linear equation of one variable (一元线性方程的图解) Suppose that there is a linear relationship between a group of experimental data variables: $y = a + bx$. The slope and intercept of the equation are b and a. Two points of $a_1(x_1, y_1)$, $a_2(x_2, y_2)$ with appropriate distance are chosen on the line, as shown in Fig. 3-11. Through the graphic, the slope of the straight line is $b = \dfrac{y_2 - y_1}{x_2 - x_1}$. If the origin of the coordinate axis is 0, the intercept of the straight line can be directly read on the y-axis (where, $x = 0$, $y = a$). Another way is to extend the line toward the longitudinal axis by extrapolation, the intersect point c is the intercept of the straight line. Otherwise, the intercept of the line is calculated by the following formula:

$$a = \frac{y_1 x_2 - y_2 x_1}{x_2 - x_1} \qquad (3-16)$$

$a_1(x_1, y_1)$ and $a_2(x_2, y_2)$ are any two points on the straight line. In order to obtain the

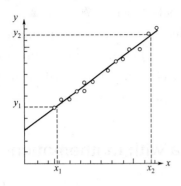

Fig. 3-11　Graphical representation of a linear equation of one variable

图 3-11　一元线性方程的图解

maximum accuracy, it is appropriate to select the points as far as possible with integral value on the line.

If the slope is diagrammatically applied on the log coordinates, note the correct method of the slope.

② Graphical representation of a binary linear equation（二元线性方程的图解）　If the studied physical quantity, i.e. the dependent variable, is linear with two variables, which can be expressed as the following function:

$$y = a + bx_1 + cx_2 \tag{3-17}$$

The above equation is a binary linear equation function, the constants a, b and c in the formular can be determined by graphical method. The first step is to keep one of the variables constant, e.g., regarding x_1 as a constant. The upper equation can be rewritten as $y = d + cx_2$, where $d = a + bx_1 =$ constant.

A straight line can be plotted from the data of y and x_2 in the rectangular coordinate as shown in Fig. 3-12. The coefficient c of x_2 is determined by the above graphical method. For any two points $e_1(x_{21}, y_1)$ and $e_2(x_{22}, y_2)$ on the line in Fig. 3-12(a), there is:

$$c = \frac{y_2 - y_1}{x_{22} - x_{21}} \tag{3-18}$$

Substitution of obtained c into the original formula and rewritten as follows:

$$y - cx_2 = a + bx_1 \tag{3-19}$$

Let $y' = y - cx_2$

$$y' = a + bx_1 \tag{3-20}$$

y' is given by c and experimental data y and x_2. After plotting the straight line in Fig 3-12(b) according to y' and x_1, any two points $f_1(x_{11}, y'_1)$, $f_2(x_{12}, y'_2)$ are taken on the line. Then constants a and b can be determined from two points f_1 and f_2:

$$b = \frac{y'_2 - y'_1}{x_{12} - x_{11}} \tag{3-21}$$

$$a = \frac{y'_1 x_{12} - y'_2 x_{11}}{x_{12} - x_{11}} \tag{3-22}$$

When a and b are determined, the independent variables x_1, x_2 should be changed accordingly, so that the results can cover the whole experimental range.

(2) Method of selected points（选点法）

The method of selected points is also called the equation solving approach, which is applicable to conditions with high accuracy of experimental data. Otherwise the obtained function will be meaningless. The specific steps are:

① Selection of appropriate empirical equation, $y = f(x)$.

② Establishment of equation set for undetermined constants. If the empirical equation is selected as: $y = a + bx$, two experimental points (x_1, y_1), (x_2, y_2) are selected from the experimental data and substituted into the above equation:

Chapter 3 Methods of Processing Experimental Data

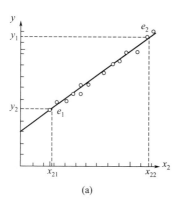 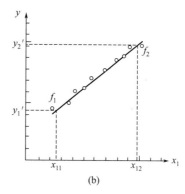

Fig. 3-12 Graphical representation of a binary linear equation

图 3-12 二元线性方程的图解

$$\begin{cases} a+bx_1=y_1 \\ a+bx_2=y_2 \end{cases} \tag{3-23}$$

③ The constants a and b can be obtained by solving the above equation.

The method of selected points can also combine with the graphic method. First, the experimental data are plotted on the coordinates, and a straight line representing all the data is drawn using a ruler. The experimental points on both sides of the line should be evenly distributed and close to the straight line. Two points at both ends of the line are substituted into the empirical formula, and then the constant can be obtained by solving simultaneous equations.

(3) Mean value method (平均值法)

If the function is linear or linearizable, the function fits $Y=A+BX$. The conditional equations $Y_i = A + BX_i$ are listed, where the number of equations is equal to the number of known experiments. Adding the conditional equations together in accordance with the principle of equality of even numbers or approximate equality of odd numbers, the following two equations are obtained:

$$\sum_1^m Y_i = mA + B\sum_1^m X_i \tag{3-24}$$

$$\sum_{m+1}^n Y_i = (n-m)A + B\sum_{m+1}^n X_i \tag{3-25}$$

The coefficients A and B can be obtained by solving the above equations.

Example: A set of data of Re and $Nu/Pr^{0.4}$ from heat transfer experiments is shown in Table 3-3.

Table 3-3 A set data of Re and $Nu/Pr^{0.4}$

表 3-3 一组 Re 和 $Nu/Pr^{0.4}$ 相关数据

Re	4.25×10^4	3.72×10^4	3.45×10^4	3.18×10^4	2.56×10^4	2.14×10^4
$Nu/Pr^{0.4}$	86.7	82.1	78.0	70.0	61.2	53.9

Its empirical equation is as follows:

$$\frac{Nu}{Pr^{0.4}}=ARe^n \tag{3-26}$$

Try to determine the coefficients A and n by the mean value method.

Solution: After taking logarithm of the empirical formula and linearizing it, we can get:

$$\lg\left(\frac{Nu}{Pr^{0.4}}\right)=\lg A+n\lg Re \tag{3-27}$$

Taking the logarithm of above experimental data, we can get Table 3-4:

Table 3-4　$\lg Re$ and $\lg(Nu/Pr^{0.4})$

表 3-4　$\lg Re$ 和 $\lg(Nu/Pr^{0.4})$

$\lg Re$	4.6284	4.5705	4.5378	4.5024	4.4082	4.3324
$\lg(Nu/Pr^{0.4})$	1.9380	1.9143	1.8921	1.8451	1.7868	1.7316

Divide the data into two equal groups and then add them, we can get Table 3-5.

Table 3-5　The results of the addition

表 3-5　加法运算结果

$1.9380=\lg A+4.6284B$ $1.9143=\lg A+4.5705B$ $1.8921=\lg A+4.5378B$	$1.8451=\lg A+4.5024B$ $1.7868=\lg A+4.4082B$ $1.7316=\lg A+4.3324B$
$5.7444=3\lg A+13.7367B$	$5.3635=3\lg A+13.2430B$

Solve the following equation set:

$$\begin{cases}5.7444=3\lg A+13.7367B\\5.3635=3\lg A+13.2430B\end{cases} \tag{3-28}$$

Gives

　　$B=0.77$, and $A=0.024$

Therefore, the dimensionless number equation is

$$\frac{Nu}{Pr^{0.4}}=0.024Re^{0.77} \tag{3-29}$$

　　(4) The method of least squares（最小二乘法）

In graphic solution, there will be errors in drawing points on the coordinate paper, and it is artificial to determine the position of a straight line according to the distribution of points. Therefore, it is often not accurate to determine the slope and intercept of a line by graphical method. The more accurate method is the least square method, which assumes that the best-fit curve of a given type is the curve that can minimize the sum of squares of the deviations between each data point and the regression equation（最佳的直线就是能使各数据点同回归线方程求出值的偏差的平方和为最小）. The following is a detailed derivation of the mathematical expression.

　　① Unary linear regression（一元线性回归）　There are n experimental data points: $(x_1,y_1),(x_2,y_2),\cdots,(x_n,y_n)$.

Suppose that the optimal linear function relation is $y=b_0+b_1x$, then n sets of y'

values corresponding to each x value can be calculated based on this formula.

$$y'_1 = b_0 + b_1 x_1$$
$$y'_2 = b_0 + b_2 x_2$$
$$\cdots$$
$$y'_n = b_0 + b_n x_n$$

However, the measured values for each x are y_1, y_2, $\cdots$, y_n, so each set of experimental values has the deviation (error) δ from the corresponding calculated value y':

$$\delta_1 = y_1 - y'_1 = y_1 - (b_0 + b_1 x_1)$$
$$\delta_2 = y_2 - y'_2 = y_2 - (b_0 + b_2 x_2)$$
$$\cdots$$
$$\delta_n = y_n - y'_n = y_n - (b_0 + b_n x_n)$$

According to the principle of least square method, the sum of squares of the deviation between the measured value and the true value is the smallest.

The necessary condition for $\sum_{i=1}^{n} \delta_i^2$ to be minimal is:

$$\begin{cases} \dfrac{\partial \left(\sum\limits_{i=1}^{0} \delta_i^2 \right)}{\partial b_0} = 0 \\ \\ \dfrac{\partial \left(\sum\limits_{i=1}^{n} \delta_i^2 \right)}{\partial b_1} = 0 \end{cases} \quad (3\text{-}30)$$

Rearranging it gives

$$\frac{\partial (\sum \delta_i^2)}{\partial b_0} = -2[y_1 - (b_0 + b_1 x_1)] - 2[y_2 - (b_0 + b_2 x_2)] - \cdots - 2[y_n - (b_0 + b_n x_n)] = 0$$

$$\frac{\partial (\sum \delta_i^2)}{\partial b_1} = -2x_1[y_1 - (b_0 + b_1 x_1)] - 2x_2[y_2 - (b_0 + b_2 x_2)] - \cdots - 2x_n[y_n - (b_0 + b_n x_n)] = 0$$

The sum form is

$$\begin{cases} \sum\limits_{i=1}^{n} y - nb_0 - b_0 \sum\limits_{i=1}^{n} x = 0 \\ \\ \sum\limits_{i=1}^{n} xy - b_0 \sum\limits_{i=1}^{n} x - b_1 \sum\limits_{i=1}^{n} x^2 = 0 \end{cases}$$

By solving the equations:

$$\begin{cases} b_0 = \dfrac{\sum\limits_{i=1}^{n} x_i y_i \sum\limits_{i=1}^{n} x_i - \sum\limits_{i=1}^{n} y_i \sum\limits_{i=1}^{n} x_i^2}{\left(\sum\limits_{i=1}^{n} x_i \right)^2 - n \sum\limits_{i=1}^{n} x_i^2} \\ \\ b_1 = \dfrac{\sum\limits_{i=1}^{n} x_i \sum\limits_{i=1}^{n} y_i - \sum\limits_{i=1}^{n} x y_i}{\left(\sum\limits_{i=1}^{n} x_i^2 \right) - n \sum\limits_{i=1}^{n} x_i^2} \end{cases} \quad (3\text{-}31)$$

The resulting linear equation with an intercept of b_0 and a slope of b_1 is the best associated line for each experimental point.

② Correlation coefficient—significance test of linear relation（线性关系的显著性——相关系数）

After we we've figured out how to regression line, there is still the problem of testing whether it makes sense. A statistical measure called correlation coefficient (r) is introduced to determine the degree of linear correlation between two variables.

$$r = \frac{\sum_{i=1}^{n}(x-\bar{x})(y-\bar{y})}{\sqrt{\sum_{i=1}^{n}(x-\bar{x})^2 \sum_{i=1}^{n}(y-\bar{y})^2}} \tag{3-32}$$

Where,
$$\bar{x} = \frac{1}{n}\sum_{i=1}^{n}x_i$$
$$\bar{y} = \frac{1}{n}\sum_{i=1}^{n}y_i$$

In probability, it can be proved that the absolute value of the correlation coefficient of any two random variables is not greater than 1, namely

$$|r| \leqslant 1 \text{ or } 0 \leqslant |r| \leqslant 1$$

The physical significance of r is the linear correlation degree of two random variables x and y（r 的物理意义是表示两个随机变量 x 和 y 的线性相关的程度）, which are illustrated in several cases：

When $r = \pm 1$, n groups of experimental values all fall on the straight-line $y' = b_0 + b_1 x$, and is called complete correlation.

As the closer the $|r|$ approaches to 1, that is the closer the n groups of experimental value (x_i, y_i) are to the straight line $y' = b_0 + b_1 x$, the relation between the variables y and x is more linear.

When $r = 0$, there is completely no linear relationship between the variables. It should be noted that when r is very small, it is not linear, but not equal that there is no other relationship.

③ Binary linear regression（二元线性回归） The regression analysis of multiple factors affecting the dependent variable is multiple regression analysis（多元回归分析）. The principle of multiple linear regression is exactly the same as that of unary linear regression, but it is much more computationally complex. The binary regression method is introduced below.

If the function $y = b_0 + b_1 x_1 + b_2 x_2$ meet the linear relation, n groups of x_1, x_2, y values are obtained from the experiment. The experimental data are correlated according to the least squares principle to solve the determined coefficients. A simple derivation of the mathematical expression are as follow.

Let the optimal linear function relationship be $y' = b_0 + b_1 x_1 + b_2 x_2$, the corresponding y' can be calculated from n groups of x_1, x_2. Then,

$$y'_1 = b_0 + b_1 x_{11} + b_2 x_{21}$$
$$y'_2 = b_0 + b_1 x_{12} + b_2 x_{22}$$
$$\cdots$$
$$y'_n = b_0 + b_1 x_{1n} + b_2 x_{2n}$$

When measured, the deviation of each experimental value and the corresponding calculated value y' is

$$\delta_1 = y_1 - y'_1 = y_1 - (b_0 + b_1 x_{11} + b_2 x_{21})$$
$$\delta_2 = y_2 - y'_2 = y_2 - (b_0 + b_1 x_{12} + b_2 x_{22})$$
$$\cdots$$
$$\delta_n = y_n - y'_n = y_n - (b_0 + b_1 x_{1n} + b_2 x_{2n})$$

According to the least squares principle（最小二乘法原理）, the sum of square of deviation between the measured value and the true value is the minimum, the necessary condition that $\sum_{i=1}^{n} \delta_1^2$ is the minimum is

$$\begin{cases} \dfrac{\partial (\sum_{i=1}^{n} \delta_i^2)}{\partial b_2} = 0 \\ \dfrac{\partial (\sum_{i=1}^{n} \delta_i^2)}{\partial b_1} = 0 \\ \dfrac{\partial (\sum_{i=1}^{n} \delta_i^2)}{\partial b_0} = 0 \end{cases}$$

Arrange the equations and get:

$$nb_0 + b_1 \sum_{i=1}^{n} x_{1i} + b_2 \sum_{i=2}^{n} x_{2i} - \sum_{i=1}^{n} y_i = 0 \tag{3-33}$$

$$b_0 \sum_{i=1}^{n} x_{1i} + b_1 \sum_{i=1}^{n} x_{1i}^2 + b_2 \sum_{i=2}^{n} x_{1i} x_{2i} - \sum_{i=1}^{n} x_{1i} y_{1i} = 0 \tag{3-34}$$

$$b_0 \sum_{i=1}^{n} x_{2i} + b_1 \sum_{i=1}^{n} x_{1i} x_{2i} + b_2 \sum_{i=2}^{n} x_{2i}^2 - \sum_{i=1}^{n} x_{2i} y_{1i} = 0 \tag{3-35}$$

By solving Eqs. (3-33), (3-34) and (3-35) simultaneously, b_0, b_1, b_2 can be obtained, consequently, the corresponding quasi-number correlation is obtained.

(5) Numerical calculation method（数值计算方法）

In chemical engineering, in addition to data regression and fitting（回归与拟合）, the numerical calculation of definite integral is often encountered. For the following integral formular

$$I = \int_{y_1}^{x_2} f(x) \mathrm{d}x \tag{3-36}$$

For the calculation of definite integral, the approximate value is generally obtained using the graph integral and numerical calculation methods.

① Graphic integration (图解积分法) Based on the function relation of Eq. (3-36), graphic integration can be carried out. For example, when the equilibrium line is a curve in the absorption process, the number of mass transfer units (传质单元数) is generally calculated by graphical integration:

$$N_{OG} = \int_{Y_2}^{Y_1} \frac{dY}{Y-Y^*} \tag{3-37}$$

The steps of the graphic integration method

a. Find $Y-Y^*$ corresponding to Y from the operation line and equilibrium line, as shown in Fig. 3-13(a);

b. Plot the curve of $Y = \left[\dfrac{1}{Y-Y^*}\right]$ in the range from Y_1 to Y_2, as shown in Fig. 3-13(b);

c. In the range from Y_1 to Y_2, the area enclosed by the curve of $Y = \left[\dfrac{1}{Y-Y^*}\right]$ and abscissa is the number of mass transfer units, as shown in the shaded section of Fig. 3-13(b).

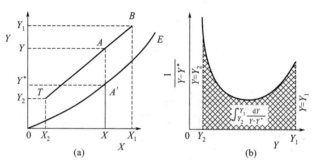

Fig. 3-13 Illustration of solving N_{OG} by graphical integration

图 3-13 图解积分法求 N_{OG}

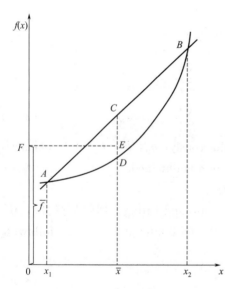

Fig. 3-14 Simpson integral method

图 3-14 辛普森积分法

② Simpson integral method (辛普森积分法) Simpson integral method is commonly used in engineering calculation, and its integral steps are as follows:

a. As shown in Fig. 3-14, plot a smooth curve on coordinate paper according to the given function relations.

b. Mark the upper and lower limits of the integral on abscissa, and the corresponding two points $A[x_1, f(x_1)]$ and $B[x_2, f(x_2)]$ on the curve. Then, connect the two points A and B into a straight line.

c. Find the median of the integral upper limit x_1 and the lower limit x_2, namely:

$$\bar{x} = \frac{x_1 + x_2}{2}$$

Draw a vertical line upward from the abscissa, and intersect the curve and line AB at two points C and D respectively.

d. Divide the line CD into three equal parts. Point E is located at 1/3DC upward from point D, and draw a horizontal line from point E to y-axis, intersecting y-axis at point F. The distance of point F from the origin is $\overline{f}$, and $\overline{f}$ is the mean value $f(x)$ at x_1-x_2. If the curve rises upwards, point D on the curve takes 1/3 down as point E.

e. Finally, the integral value is

$$I = \int_{x_1}^{x_2} f(x) \, dx = (x_2 - x_1) \overline{f} \tag{3-38}$$

Based on the principle of segmentation, the integral interval $[x_0, x_n]$ is divided into n equal parts, and the distance of each equal part is

$$h = \frac{x_n - x_0}{n} \tag{3-39}$$

where, $x_n > x_0$; n is even.

The Simpson formula for calculating the definite integral approximation is

$$I = \frac{h}{3} \left\{ f(x_0) + 4 \sum_{k=0}^{n-1} f[x_0 + (2k+1)h] + 2 \sum_{k=0}^{n-1} f(x_0 + 2kh) + f(x_n) \right\} \tag{3-40}$$

Obviously, the larger the n value, the more the interval, the more accurate the results. However, as the n value increases, the computational amount will also increase. Therefore, the appropriate value of n depends on the accuracy allowed by the result.

Computing with Simpson's formula can obtain the more exact fixed integral values, and the computer-assisted program calculation makes it more convenient and reliable.

Chapter 4　Common Measuring Instrument in Laboratory

第4章　实验室常用测量仪表

Liquid pressure, flow and temperature are important parameters necessary to measure and control in scientific experiments and chemical production. Suitable measuring instruments should be selected according to the needs of the experiment, which can be purchased directly from the market or self-designed. To achieve reasonable use, it is necessary to have a preliminary understanding of the measuring instrument. The following is a brief introduction of the principle, characteristics and application of the instruments used in pressure, flow and temperature measurement.

4.1　Pressure measurement
（压力测量）

Pressure measurement is often encountered in chemical industry and scientific experiments. For example, it is necessary to measure the pressure of the tower top and kettle in the chemical units of distillation and absorption to understand whether the tower operation is normal. The pressure drop of the fluid flowing through the pipeline in the test of pipeline resistance measurement and the pump inlet and outlet pressure in the test of pump characteristic curve measurement are all essential and important parameters.

Pressure measuring instruments can be roughly divided into: liquid column pressure gauge（液柱式压力计）, elastic pressure gauge（弹性式压力计）and electrical type pressure gauge（电气式压力计）, etc.

(1) Liquid column pressure gauge（液柱式压力计）

Liquid column pressure gauge (both shaping products and self-made) is a pressure measuring instrument made by the principle that the pressure generated by the height of the liquid column is balanced with the measured pressure（利用液柱高度产生的压力和被测压力相平衡的原理制成的测压仪表）, which has the characteristics of simple structures, easy to use, high precision and low price. It is widely used in industrial production and laboratory for measuring low pressure or vacuum degree.

① Structure of the liquid column pressure gauge　There are three structural forms of liquid column pressure gauge: U-tube manometer (U 形管压力计), single tube manometer (单管压力计, also called cup manometer) and inclined manometer (斜管压力计). The structure form and characteristics of the liquid column manometer are shown in Table 4-1.

Table 4-1　Structure form and characteristics of liquid column manometer
表 4-1　液柱式压力计的结构形式和特性

Name	Schematic diagram	Measuring range	Static equation	Note
U-tube manometer (U 形管压差计)		$R \leqslant 800mm$	$\Delta p = Rg(\rho_o - \rho)$ (liquid) $\Delta p = Rg\rho$ (gas)	No need for zero adjustment before use, often used as a standard differential pressure technique to calibrate the flow meter
Inverted U-tube manometer (倒置 U 形管压差计)		$R \leqslant 800mm$	$\Delta p = Rg(\rho_o - \rho)$ (liquid)	Take the liquid to be measured as the indicator, suitable for the measurement of small pressure difference
Single tube manometer (单管压差计)		$R \leqslant 1500mm$	$\Delta p = R\rho g(1 + S_1/S_2)$ When $S_1 \ll S_2$ $\Delta p = R\rho g$ S_1, S_2: Cross-sectional area of vertical pipe or the enlarged chamber (the same below)	Zero point is at the lower end of the ruler, need to adjust the zero point before use, and can be used as etalon
Inclined manometer (斜管压差计)		$R \leqslant 200mm$	$\Delta p = L\rho g(\sin\alpha + S_1/S_2)$ When $S_1 \gg S_2$, $\Delta p = L\rho g \sin\alpha$	When α is less than 10°-20°, the measurement range can be adjusted by changing the magnitude of α. Zero point is at the lower end of the ruler and needs to be adjusted before use
Two-fluid U tube manometer (U 形管双指示液压差计)		$R \leqslant 500mm$	$\Delta p = Rg(\rho_1 - \rho_2)$	Designed to improve measurement accuracy. Equipped with two indicating fluids of similar density, and with "expansion chambers" above both arms

The structure of the U-tube manometer is shown in Table 4-1. A glass tube with an inner diameter of 6-10mm is bent into a U-shape and then fixed vertically on the plate. The scale ruler is mounted between the arms with zero point in the center of the scale. The pipe is filled with water, mercury or other liquid at a level consistent with the zero scale. The liquid height in two tubes should be recorded when measuring the pressure difference of liquid with a U-tube manometer（用U形管压力计测量液体的压力差时，必须读出两管中液面的高度）.

To measure the pressure difference between two sections of the fluid flowing through the horizontal pipe, according to the basic equation of hydrostatics, we can get:

$$\Delta p = R(\rho_o - \rho)g \tag{4-1}$$

Where, Δp—the pressure difference between the two sections of the pipe, Pa;

R—height reading of liquid column in U-tube manometer, m;

ρ_o—density of indicating liquid of U-tube manometer, kg/m^3;

ρ—the density of the liquid to be measured, kg/m^3.

The U-tube manometer can be used to measure the pressure difference of the fluid, as well as the pressure at any point of the fluid. If one end of the U-shaped tube is connected to a certain section of the equipment or pipe and the other end connects to the atmosphere, the reading R from the U-shaped tube manometer reflects the difference between the absolute pressure of the fluid in a section of the equipment or pipeline and the atmospheric pressure, namely, the gauge pressure（如果U形管的一端与设备或管道某一截面相连，另一端与大气相通，这时从U形管压力计的读数R是反映设备或管道中某截面流体的绝对压强与大气压强之差，即为表压强）.

② Notes in use of liquid column pressure gauge（液柱式压力计使用注意事项）Although the liquid column pressure gauge has the advantages of simple structure, convenient use and high measurement accuracy, its pressure resistance is poor, the structure is easy to break, and the measurement range correction and indicated value are related to the density of the working liquid. So the following points must be noted in use:

a. The measured pressure must not exceed the measuring range of the instrument. Special attention is that the pressure will increase due to the sudden pressurization of the tested object or careless operation causing the indicator fluid to outrush.

b. Avoid installation in places with overheating, undercooling, corrosive liquids, or vibration.

c. Note that the selected indication liquid does not miscible or react with the tested liquid. Select the appropriate indication liquid according to the measured pressure. Commonly used indicating liquids are mercury, water, carbon tetrachloride, benzyl alcohol, kerosene, glycerol, etc.（根据所测的压力大小，选择合适的指示液体，常用指示液体如水银、水、四氯化碳、苯甲醇、煤油、甘油等）. When filling the indicating liquid, align the liquid level to the zero point of scale.

d. Because of the capillary phenomenon of the liquid (Fig. 4-1), when reading the pressure value, the sightline should focus on the surface of liquid column. If the level (water) is concave, the reading should be level with the bottom of the concave, and if the level (mercury) is convex, the reading should be level with the top of the convex. During use, the measuring tube and scale should be kept clear, the working fluid should be replaced regularly. It should also be checked frequently for leaks between the instrument and the connecting pipe.

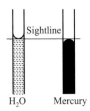

Fig. 4-1 Capillarity of water and mercury in glass tubes

图 4-1 水和水银在玻璃管中的毛细现象

(2) Elastic manometer (弹性压力计)

Elastic pressure gauge is a pressure instrument based on the principle of elastic deformation produced by various elastic pressure sensors of different shapes under the action of measured pressure. This instrument has the advantages of simple structure, firm and reliable, wide range of pressure measurement, convenient use, low cost and sufficient accuracy, etc., and is easy to be made into signal transmitting, remote indication and control unit. So, it is the most widely used pressure gauge in the industry.

Elastic manometers have different elastic elements according to the size of the pressure measurement range. According to the shape and structure of the elastic elements, the elastic manometers are in four forms: Bourdon gauges (single-coil bourdon pressure gauges, multi-coil bourdon pressure gauges) [弹簧压力表（单圈弹簧管压力表、多圈弹簧管压力表）], diaphragm pressure gauges (膜片压力表), capsule pressure gauge (膜盒压力表) and bellow pressure gauge (波纹管压力表).

① Bourdon gauges (弹簧压力表) Bourdon gauge is divided into single-coil spring tube pressure gauge and multi-coil bourdon pressure gauge. Single-coil spring pipe pressure gauge with various varieties and models can be widely used for vacuum measurement, or for high-pressure measurement up to 10^3 MPa. According to the range of pressure measurement, it is generally divided into pressure gauge (压力表), vacuum gauge (真空表) and pressure vacuum gauge (压力真空表). According to the precision grade, there are precision pressure gauges (precision grade 0.25), standard pressure gauges (precision grade 0.4) and ordinary pressure gauges (precision grade 1.5 and 2.5). According to the use, there are pressure gauges, vacuum gauges, ammonia pressure gauges, oxygen pressure gauges, acetylene pressure gauges, hydrogen pressure gauges etc. According to the signal display mode, it is divided into double needle double tube pressure gauge (that is, two single tube pressure measuring system installed in a case, can measure two pressures), electric contact pressure gauge, remote transmission pressure gauge, etc. According to the adaptability of the special environment, there are explosion-proof pressure gauges, vibration-proof pressure gauges, sulfur-proof pressure gauges, acid-proof pressure gauges, etc. Multi-coil bourdon pressure gauges have high sensitivity and are commonly used in pressure thermometers.

The structure of ordinary single-coil bourdon gauge is shown in Fig. 4-2.

The measured pressure enters by the joint 8, forcing the free end of the spring tube 1 to expand to the upper right. The elastic deformation displacement of the free end makes the gear 3 deflect counterclockwise through the pull rod 2, the pointer 5 then rotates clockwise

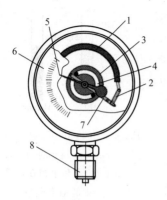

Fig. 4-2　Bourdon gauges
图 4-2　弹簧压力表
1—Spring tube（弹簧管）；2—Pull rod（拉杆）；
3—Gear（齿轮）；4—Center gear（中心齿轮）；
5—Pointer（指针）；6—Dial（刻度盘）；
7—Drive mechanism（传动机构）；
8—Joint（接头）

driven by the coaxial center gear 4, so that the measured pressure is displayed on the dial 6 of the panel. One end of the hairspring is fixed to the central gear shaft and the other end is on the bracket. With the help of the spring force, the central gear and the sector gear are always engaged on only one side of the meshing surface, so that the measurement error caused by the inherent meshing gap between the sector gear and the central gear can be eliminated. The end where the gear connect with the pull rod has an open slot. The transmission mechanism can be controlled by changing the connection position of the pull rod and the sector gear.

To ensure the measurement accuracy and long-term use of the bourdon gauge, the installation and maintenance of the instrument are very important, attention should be paid to the following provisions during use.

a. Use the instrument within the normal allowable pressure range. The general pressure should not exceed 70% of the upper limit of measurement, and 60% of the measured limit during pressure fluctuation. Industrial pressure gauges should be used under the conditions of ambient temperature $-40\text{-}60℃$ and relative humidity not greater than 80%.

b. The distance between the instrument installation place and the pressure measuring point is as short as possible to avoid slow indication. Moreover, the installation height of the instrument shall be the same or close to the pressure measuring point, otherwise the additional pressure error of the liquid column will occur, and the correction value is added if necessary.

c. Using instruments under vibration, install vibration absorber device, and install isolator when measuring medium with crystallization or large viscosity. The instrument must be installed vertically without leakage. The cut-off valve should be installed between the pressure taking port and the pressure gauge for repairing the pressure gauge.

d. Use special instruments such as oxygen pressure gauge when measuring the pressure of explosion, corrosion and toxic gas, and no contact with oil to avoid explosion（测量爆炸、腐蚀、有毒气体的压力时，应使用特殊的仪表，如氧气压力表，严禁接触油类，以免发生爆炸）.

e. Check the instrument regularly, and use the qualified instrument.

② Diaphragm pressure gauge（膜片压力表）　The greatest advantage of the diaphragm pressure gauge is available to measure medium pressure with high viscosity. If the diaphragm and the lower cover are made of stainless steel, or if the inner side of the diaphragm and the lower cover is coated with an appropriate protective coating（e.g., fluoroplastic）, it can also be used to measure the pressure of certain corrosive media.

③ Capsule pressure gauge（膜盒压力表） Capsule pressure gauge is suitable for measuring air, or micro-and negative pressure of gases not corrosive to copper alloys.

④ Bellow pressure gauge（波纹压力表） Bellow pressure gauges are commonly used to measure pressure in brass and carbon steels from 0-400kPa in non-corrosive, low viscosity, clean, non-crystalline and non-solidifying media. Because the bellows shift greatly under the action of pressure, it is generally made into an automatic recording instrument except for the indicator instrument. Some bellow pressure gauges also have electrical contact devices and regulating devices.

⑤ Electrical pressure gauge（电气式压力表） In order to adapt to the modern industrial production process to carry on the long-distance transmission, the display, the alarm, the detection and the automatic adjustment to the pressure measurement signal as well as to facilitate the application of the computer technology and so on, the electric type pressure gauge is often used.

An electric manometer is an instrument that converts pressure values into electricity. It is generally composed of pressure sensor, measuring circuit and indicating and recording device.

Most pressure sensors still use elastic elements as pressure sensing elements. The displacement of the elastic element under the action of pressure transforms into a certain electric quantity by the electrical device. This power is then measured by corresponding instrument (called the secondary instrument) and expressed by the pressure value［弹性元件在压力作用下的位移通过电气装置转变为某一电量，再由相应的仪表（称二次仪表）将这一电量测出，并以压力值表示出来］. Such electrical pressure gauges include resistance type, inductance type, capacitive type, Hall type, strain type and vibration string type, etc. Another category is the electrical pressure gauge made of the physical properties related to pressure of some objects, such as pressure sensors made of piezoelectric crystal and piezoresistor. Such pressure sensor itself can generate far-transmitted electrical signals.

（3）Selection of pressure point and pressure tap（取压点的选择及取压孔）

① Selection of pressure point To measure the static pressure correctly, it is very important to choose the pressure measuring point. The pressure measuring point must be selected at the minimum fluid flow interference. For example, to measure the pressure on the pipeline, the pressure measuring point should be chosen at the distance of 40-50 times the inner diameter of the pipe, fittings or other obstruction upstream of the fluid. After the turbulent streamline flows through the stable section for a distance, the streamline near the wall surface is parallel to the wall surface to avoid the influence of kinetic energy on the measurement. If the stable section within $(40\text{-}50)d_{in}$ cannot be guaranteed due to conditions, rectifying plate or rectifying tube can be set up.

② Pressure tap（取压孔口） Pressure tap (also known as pressure measuring hole) is connected to the pressure gauge or pressure instrument by the pressure pipe (also called as piezometer tube) to show the pressure at the pressure measuring point. Since the orifice opens on the pipe wall, the flow line bends into the orifice as the fluid flows through the orifice and causes the vortex. Therefore, there is an error between the static pressure from pressure outlet and the real static pressure of the fluid. This error relates to the flow state near the orifice, the size geometry and depth of the orifice, the direction of the orifice axis and the roughness of the wall surface at the opening. The larger the aperture size is, the more serious the streamline bending will be, resulting in the larger the eddy and the greater the measurement error. Therefore, theoretically, the pressure tap should be as small as possible, but the orifice is too small for processing, and it is easy to get clogged up with dirt. In addition, the measured dynamic performance is poor. General aperture is 0.5-1mm (with lower accuracy requirements, the aperture size can be properly enlarged). The hole depth h/aperture size $d \geqslant 3$, the axis of the aperture should be perpendicular to the wall surface, the edge of the hole should not be burr, the pipe wall around the hole should be smooth, there should be no concave and convex parts. The orientation of the pressure aperture depends on the specific situation of the fluid: The orifice is generally located above the pipeline for the case of gas, while on the side of the pipeline for the case of steam. For liquid, the orifice is located at an angle of 45° to the horizontal axis as shown in Fig. 4-3.

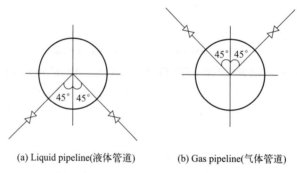

(a) Liquid pipeline(液体管道)　　(b) Gas pipeline(气体管道)

Fig. 4-3　Pressure tap of the fluid pipe

图 4-3　所示流体管道的取压口

Since the static pressure at the section is represented by the measured pressure value on the pipe wall, a pressure ring (as shown in Fig. 4-4) can be installed on this section instead of a single aperture to eliminate the additional errors caused by the uneven flow of static pressure difference at each point on the pipeline section.

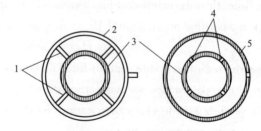

Fig. 4-4　Pressure ring

图 4-4　取压环

1—Piezometer tube（取压管）；2—Annular pipe（环形管）；3—Pipeline（管道）；
4—Pressure tap（取压孔口）；5—Pressure ring（取压环）

4.2　Metering of fluids
（流量测量）

Flow rate is an important parameter in the production process and scientific experiment of the chemical industry. In chemical and scientific experiments, the flow should be measured in order to calculate the conveying and proportioning of materials. The process flow of the flow medium, the balance of materials and energy closely relate to the flow, and the automation and optimization control of industrial production is even more inseparable from the measurement and control of the flow.

Flow rate refers to the mass of the fluid or the volume of the fluid per unit time. The former (kg/h) is called mass flow rate and the latter (m^3/h) is called volume flow rate. There are many methods and instruments for measuring flow. According to the action principle, the commonly used flow measurement instruments in industry are classified as: Area flowmeter, differential pressure flowmeter, current flowmeter and volumetric flowmeter（面积式流量计、压差式流量计、流速式流量计和容积式流量计）, etc. The flow measurement range, accuracy, application occasions and related characteristics of corresponding instrument products are shown in Table 4-2.

(1) Area type flowmeter（面积式流量计）-Rotameter（转子流量计）

① The structure of rotameter　Rotameter is also called the area meters. Because the pressure difference at both ends of the float is always constant when the float is in any balance, the rotameter is also known as constant differential pressure flowmeter（恒压差式流量计）.

Rotameter has a very extensive application. It is divided into three series of glass rotameter, pneumatic remote rotameter and electronic remote rotameter（玻璃管流量计、气远传转子流量计和电远传转子流量计）. Among them, the glass rotameter is used for the on-site measurement of transparent fluid medium. The latter two rotameters can observe the flow rate farther away from the field or from the secondary display instrument through gas signal or electrical signal. The fluid medium of this time is not necessarily required to be transparent. However, whatever the rotameter, the measuring principle is the same.

The structure of the glass rotameter is shown in Fig. 4-5. Glass rotameter is mainly composed of three parts: supporting connector, tapered tube and float（支承连接件、锥管、浮子）.

a. Supporting connector: According to different models and caliber, it is divided into flange connection, threaded connection and hose connection.

b. Tapered tube: Usually made of boron hard glass, but also made of plexiglass.

c. Float（转子）is usually in two shapes. Fig. 4-6(a) is commonly used for other flow measurements, while Fig. 4-6(b) is mostly used for large fluid flow measurements. Floats may be constructed of metals of various densities from lead to aluminum or from glass or plastic according to the nature of the measured medium and the amount of the measured flow. Stainless-steel floats are common. Floats can be made into hollow, and some into solid.

Table 4-2　Classification table of flowmeter
表 4-2　流量计分类

	Instrument	Measuring range	Accuracy	Application occasions	Characteristics
Area type	Glass rotameter (玻璃转子流量计)	16-1×10⁶ L/h(gas) 1.0-4×10⁴ L/h(liquid)	2.5	For the measurement of small flow rates of air, nitrogen, water and other similar safe fluids	① Simple structure, easy maintenance; ② Low accuracy; ③ Not suitable for toxic media and opaque media
	Metal rotameter (金属转子流量计)	0.4-3000 m³/h (gas) 12-1×10⁵ L/h(liquid)	1.5 2.5	① Situations of great flow changes; ② High viscosity, corrosive fluid; ③ Differential pressure guide pipe and easy vaporization	① Own the main characteristics of the glass tube rotor flowmeter; ② Long distance transmission; ③ With anti-corrosion, can be used for acid, alkali, salt and other corrosive media
	Plugger type flowmeter (冲塞式流量计)	4-60 m³/h	3.5	Field indication and product calculation of various no dross and no coking media	① Simple structure; ② Easy to install and use; ③ Low precision, cannot be used for pulse flow measurement
Differential pressure type	Throttle type flowmeter (节流装置流量计)	60-25000 mmH₂O	1	Flow measurement of a non-strongly corroded unidirectional fluid, allowing for a certain pressure loss	① Widely used; ② Simple structure; ③ Without individual calibration for a standard throttling device
	Constant velocity tube flowmeter (匀速管流量计)			Flow measurement of various gases and liquids with large caliber and large flow rate	① Simple structure; ② Convenient installation, disassembly and maintenance; ③ Small pressure loss, less energy consumption; ④ Low output pressure difference
Current type	Rotor type water meter (旋翼式水表)	0.045-2800 m³/h	2	Mainly used for water measurement	① Simple structure, small phenotype, high sensitivity; ② Easy installation and use

Chapter 4 Common Measuring Instrument in Laboratory

(continued)

	Instrument	Measuring range	Accuracy	Application occasions	Characteristics
Current type	Turbine flowmeter (涡轮流量计)	0.04-6000m³/h (liquid) 2.5-350m³/h (gas)	0.5-1	Used for clean fluids with low viscosity and high precision measurement in a wide measuring range	① High precision, suitable for metering; ② Wide temperature and pressure range; ③ Transmitter is small in size and easy to maintain; ④ The bearing is easy to damage, and the continuous service cycle is short
	Vortex flowmeter (旋涡流量计)	0-3m³/h (water) 0-30m³/h (gas)	1.5	Suitable for the measurement of various gases and low-viscosity liquids	① Wide range variation ability; ② Simple structure and convenient maintenance
	Electromagnetic flowmeter (电磁式流量计)	2-5000m³/h	1	Applicable to measure the flow rate of conductive liquid with electrical conductivity $>10^{-4}$S/cm	① Only for measuring conductive liquids; ② Measurement accuracy is not affected by the change of medium viscosity, density, temperature and conductivity; ③ Almost no pressure loss; ④ Not suitable for measuring ferromagnetic materials
	Split rotor steam flowmeter (分流旋翼式蒸汽流量计)	0.05-12t/h	2.5	Accurately measure the mass flow rate of saturated water vapor	① Convenient installation; ② Direct reading type, easy to use; ③ The flow of saturated water vapor can be compensated by pressure correction
Volumetric type	Oval gear flowmeter (椭圆齿轮流量计)	0.05-120m³/h	0.2-0.5	Suitable for the measurement of high-viscosity medium flow	① High accuracy; ② High metering stability; ③ Not suitable for fluids containing solid particles
	Wet gas flow meter (湿式气体流量计)	0.2-0.5m³/h		Directly used to measure gas flow, or as standard measuring instruments to calibrate other flow meters	① Measure the total gas volume with high accuracy; ② Small error in small flow; ③ Common instruments in the laboratory

141

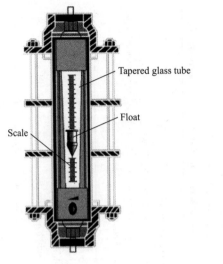
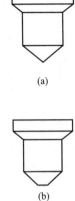

Fig. 4-5 Glass rotamer
图 4-5 玻璃转子流量计

Fig. 4-6 Float
图 4-6 转子

② The operating principles of the rotameter（转子流量计的工作原理） The rotameter consists essentially of a gradually tapered tube (usually made of glass with a taper of 40′to 3°) mounted vertically in a frame with the large end up and a freely moving float placed in the tapered tube. The fluid (gas or liquid) flows upward through the tapered tube. After flowing through the annular space between the float and the tube wall, and it flows out from the upper part of the tapered tube. As the fluid flows through the tapered tube, the float located in the tapered tube suffers an upward "impulse force" causing the float to suspend. When this force is exactly equal to the weight of the float immersed in the fluid (i.e., the weight of the float minus the buoyancy of the fluid on the float), the upper and lower forces acting on the float are in balance, and the rotor stops floating at a certain height. If the flow rate of the measured fluid suddenly increases, the "impulse force" acting on the float is increased. Due to the constant weight of the float in the fluid (i.e., the downward force acting on the float is constant), the float moves upwards. As the float moves up in the tapered tube, the annular gap between the float and the tube wall increases (i.e., the flow area increases), the flow rate of the fluid flowing through the annulus consequently decreases, so the "impulse force" decreases. When the "impulse force" is again equal to the weight of the float in the fluid, the float is stabilized at a new height. Thus, the balance position of the float in the tapered tube corresponds to the flow rate of the measured medium. If the tube is marked in divisions, the reading of the meter is obtained from the scale reading at the reading edge of the float. The above is the basic principle of the rotameter measurement.

The equilibrium position of the float in the rotameter is established by a balance of the float weight and the "impulse force" of the fluid on the float（转子流量计中转子的平衡条件是，转子在流体中的重量等于流体对转子的"冲力"）. Because the impulse force of the fluid is actually the product of the static pressure drop of the fluid between the top and bottom of the float and the cross-section area of the float, so for equilibrium:

$$V_{\text{float}}(\rho_{\text{float}}-\rho_{\text{fluid}})g=(p_{\text{upper}}-p_{\text{lower}})A_{\text{float}} \tag{4-2}$$

Where, V_{float} — volume of the float, m^3;

ρ_{float} — density of the float, kg/m^3;

ρ_{fluid} — density of the measured fluid, kg/m^3;

g — gravity acceleration, m/s^2;

p_{upper}, p_{lower} — the static pressure of the upper and lower fluid acting on the float, Pa;

A_{float} — the projected area of the float, m^2.

Since V_{float}, ρ_{float}, ρ_{fluid} and A_{float} are constant during the measurement, ($p_{\text{upper}}-p_{\text{lower}}$) should also be constant. In other words, the pressure drop of the fluid in rotameter is constant, so the rotameter measures the flow based on the method of constant pressure drop and variable throttle area(定压降变节流面积法).

Based on hydrodynamic principle, the pressure difference ($p_{\text{upper}}-p_{\text{lower}}$) can be expressed by the velocity of the fluid flowing through the annulus between the float and the tapered tube:

$$p_{\text{upper}}-p_{\text{lower}}=\xi\frac{u^2\rho_{\text{fluid}}}{2} \tag{4-3}$$

Where, ξ — resistance coefficient, related to the float shape and the viscosity of the fluid, etc., dimensionless;

u — velocity of the fluid flowing through the annulus, m/s.

From Eqs. (4-2) and (4-3), the velocity of the fluid flowing through the annulus is:

$$u=\sqrt{\frac{V_{\text{float}}(\rho_{\text{float}}-\rho_{\text{fluid}})2g}{\xi\rho_{\text{fluid}}A_{\text{float}}}} \tag{4-4}$$

If A_0 represents the cross-sectional area of the annulus between the float and the tapered tube, and $\varphi=\dfrac{1}{\xi}$ represents the correction factor, the mass flow rate of the fluid flowing through the rotameter can be calculated:

$$G=u\rho_{\text{fluid}}A_0=\varphi A_0\sqrt{\frac{2gV_{\text{float}}(\rho_{\text{float}}-\rho_{\text{fluid}})\rho_{\text{fluid}}}{A_{\text{float}}}} \tag{4-5}$$

Or it can be expressed as volumetric flow rate:

$$Q=uA_0=\varphi A_0\sqrt{\frac{2gV_{\text{float}}(\rho_{\text{float}}-\rho_{\text{fluid}})}{\rho_{\text{fluid}}A_{\text{float}}}} \tag{4-6}$$

For a given rotameter, φ is a constant. It can be seen from Eqs. (4-5) and (4-6) that the flow rate through the rotameter is only related to the annulus cross-sectional area between the float and the tapered tube. As the tapered tube gradually expands from bottom up, it relates to the height of the float. In this way, the reading of the flow rate is obtained according to the height of the float［从式（4-5）和式（4-6）可以看出，当用转子流量计来测量某种流体流量时，流过转子流量计的流量只与转子和锥形管间环隙截面积 A_0 有关。由于锥形管由下往上逐渐扩大，所以 A_0 与转子浮起的高度有关。这样，根据转子的高度就可

判断被测介质的流量大小].

③ Installation and use of rotor flowmeter（转子流量计的安装和使用）

a. The rotameter must be installed vertically and the fluid must pass through the tapered tube from bottom up. There shall be straight pipe sections of more than 5 times the pipe diameter at the inlet and outlet.

b. The instrument should be installed in a place without vibration and easy to maintain. When the instrument installing on the production pipeline, a bypass pipeline should be retrofit in parallel with the instrument to avoid affecting the normal production during maintenance. The instrument starts to run on the bypass first. When the fluid is filled in the front and back pipes of the instrument, use the instrument and turn off the bypass to avoid damage to the instrument by impact. The pipes should be cleaned before installation to prevent residual impurities from entering the instrument and affecting the normal operation.

c. The float is more sensitive to the adhesion fouling. If dirt sticks to the float, the float mass and the cross-sectional area of the annular channel will change. Sometimes it causes the float to fail to float vertically up and down, resulting in measurement errors.

d. When installing the glass pipe-type float flowmeter, the upper and lower pipes it is connected to should be fixed firmly first, must not let the instrument bear the weight of the pipe. Note that the gauge is not allowed to bear the weight of the pipe. When the measured fluid temperature is above 70℃, a protective cover shall be installed to prevent the glass rupture of the instrument when it is cold.

e. The valve shall be opened slowly when the flow rate adjusted or controlled. Otherwise, the impact caused by the sudden excessive flow rate will push the float to the top and the float would be stuck or damaged.

f. Rotameter is only suitable for measuring the clean fluid flow. When an impurity-containing fluid is measured, a filter prior to the rotameter should be installed.

④ Correction of the indicated flow value of the rotameter（转子流量计指示值修正）
The rotameter is a non-standardized instrument（非标准化仪表）, each rotor flowmeter is attached with manufacturer-calibrated flow data. For the flowmeter for measuring liquid, the manufacturer calibrates it with water at room temperature（20℃）. The flowmeter used for measuring the gas is calibrated with the air under the standard condition（标准状态，20℃，1.013×10^{5}Pa）. However, in the actual use, since the measured medium is different from the calibrated object (liquid is not water, gas is not air, density is different) and the operating state (temperature and pressure) is also different, there is a certain difference between the indicated value of the rotameter and the actual flow rate of the measured medium. Therefore, the indicated flow value must be corrected according to the density, temperature, pressure and other parameters of the measured medium.

For the liquid medium（液体介质）, the formula can be modified as follows:

$$V_{fluid} = V_0 \sqrt{\frac{(\rho_{float} - \rho_{fluid})\rho_0}{(\rho_{float} - \rho_0)\rho_1}} \tag{4-7}$$

Where, V_{fluid}—the actual flow rate of the measured medium;

V_0—the calibrated reading of the instrument with water;

ρ_{float}—the density of the float;

ρ_0—density of water used for manufacturer-calibration;

ρ_1—density of the measured medium.

For the gas medium (气体介质), the formula can be modified as follows:

$$V_{\text{fluid}} = V_0 \sqrt{\frac{\rho_0 p_0 T_1}{\rho_1 p_1 T_0}} \tag{4-8}$$

Where, V_{fluid}, ρ_1, p_1, T_1—volumetric flow rate, density, pressure and temperature of gas under operating state;

V_0, ρ_0, p_0, T_0—volumetric flow rate, density, pressure and temperature of gas under calibration state.

(2) Differential pressure flowmeter (压差式流量计)

Based on the throttling principle of fluid flow, differential pressure flowmeter uses the pressure difference generated when the fluid flows through the throttling device to achieve the flow measurement (是利用流体流经节流装置或匀速管时产生的压力差来实现流量测量的). It is composed of a throttling device and a differential pressure gauge, and is one of the most mature and commonly used flow measurement instruments in industrial production. Abundant practical experience and complete experimental data have been accumulated in the use of this flowmeter. The design and calculation of throttling devices have unified standards. Therefore, it can be manufactured and used directly from the calculated results without calibration by experimental methods. The common throttles include orifice plates (孔板), nozzles (喷嘴), Venturi pipe (文丘里管) and Venturi nozzles (文丘里喷嘴), among which the first two are the most commonly used, as shown in Fig. 4-7 and Fig. 4-8. Here, content focuses on only the throttle device and pressure gauge.

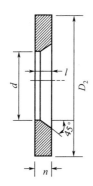

Fig. 4-7　The structure of orifice plate
图 4-7　孔板的结构

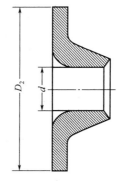

Fig. 4-8　The structure of nozzle
图 4-8　喷嘴的结构

① Throttling phenomenon and its principle (节流现象及其原理)　When the fluid stream encounters a throttle device installed in the pipe (a orifice with smaller diameter than the pipe in the middle of throttling device), the cross section suddenly decreases accompanied with increased velocity. The fluid flowing through the orifice first contracts and then expands to fill the entire pipe section, and the normal velocity distribution eventually is reestablished. The hydrostatic pressure at the pipe wall before and after the throttling orifice varies with kinetic energy, which is called throttling phenomenon（节流现象）. Therefore, when the fluid flows through the orifice at certain flow rate, there will be a certain pressure difference. The larger the flow rate, the greater the pressure difference will be. So, the measurement of the pressure difference can be used to measure the fluid flow.

The principle of differential pressure caused by fluid flows through the throttling device is called the throttling principle（流体流过节流装置产生压差的原理称为节流原理）. The relation between the pressure difference and the flow rate generated by the fluid flowing through the throttling device is:

$$V = C_0 A_0 \sqrt{\frac{2gR(\rho_0 - \rho)}{\rho}} \qquad (4-9)$$

② Commonly throttling elements and pressure measurement methods（常用的节流元件和取压方式）

a. Throttling elements（节流元件）

Orifice plate（孔板）

A standard sharp-edged orifice is shown in Fig. 4-7. It consists of an accurately machined and drilled plate with the hole concentric with the pipe in which it is mounted. The inlet cylinder portion of the standard orifice plate shall be installed concentric with the pipe. The orifice plate must be perpendicular to the pipe axis with a deviation of no more than $\pm 1°$. Orifice plate material is generally stainless steel, copper or duralumin.

Features of orifice plate: simple structure, easy to process and low cost, but the energy loss is greater than nozzle and Venturi flowmeter.

The plates shall be installed in direction and shall not be reversed（安装孔板时应注意方向，不得装反）. Machining requirements are strict, such as right-angle inlet, edge to be sharp and no burr. Otherwise, the measurement accuracy will be affected. Therefore, the orifice plate is not suitable for the dirty or corrosive media that tend to dirty, wear and deform the throttle elements during the measurement process.

Nozzle（喷嘴）

The standard nozzle is a panel with a short horn, and the inflow cross section changes gradually, as shown in Fig. 4-8. The applicable pipe diameter D for the nozzle is 50-100mm, aperture ratio is 0.32-0.8, and the Reynolds number is 2×10^4-2×10^6.

Characteristics of the nozzle: the energy loss is second only to the Venturi tube, with a higher measuring accuracy. The short straight pipe length required before and after the nozzle can be applied to corrosive, easy to wear and dirty medium measured.

Venturi tube（文丘里管）

A Venturi meter is shown in Fig. 4-9. A short conical inlet section leads to a throat section, then to a long discharge cone. The upstream inlet section has a diameter of D and cross-sectional area of F_1, followed by a contraction section with a contraction angle β of 19°-23° in general.

Chapter 4 Common Measuring Instrument in Laboratory

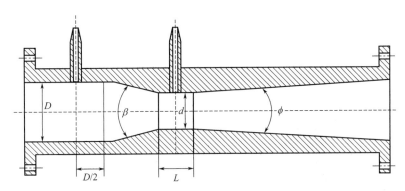

Fig. 4-9 Structure diagram of the Venturi meter
图 4-9 文丘里流量计结构图

In the middle there is a flat throat of diameter d, cross-sectional area F_2, and the length of the flat throat L is equal to d. The last part is an expansion section with an expansion angle of 5°-15°, which makes the pipe of the flowmeter gradually transition to the same size as the original pipe section.

The fluid accelerates and decompresses through the contraction section, so that the static pressure at the throat is less than that at the upstream inlet section. The greater the flow velocity, the greater the static pressure difference between the throat and the upstream section. The static pressure difference reflects the rate of flowing. The static pressure of the inlet section is p_1, and p_2 at the throat. There are smooth straight sections with a length of 8D and 5D before and after the Venturi tube. Ratio of throat section to pipe section A_1/A_2 is generally between 0.2-0.5.

Characteristics of the Venturi tube: the energy loss is the smallest of the various throttle elements, and the pressure can recover after the fluid flows through the Venturi tube. However, Venturi tube manufacturing is complex and has a high cost.

b. Pressure-tapping（取压方式） There are many pressure-taking methods of throttling device. The flow coefficient is not the same with the different pressure-taking mode. In terms of orifice plate, there are roughly four methods: the corner tapping（角接取压法）, the flange tapping（法兰取压法）, the theoretical tapping（理论取压法）and the span（$D-D/2$）tapping（径距取压法）. In particular, the corner tapping, the flange tapping are the most widely used.

The corner tapping（角接取压法）: The upstream and downstream pressure measuring ports are set on the position immediately before and after the throttle element (hole plate), this pressure measurement method is called the corner tapping. The methods of corner tapping fall into two kinds: ring chamber tapping（环室取压法）and separate bore tapping（单独钻空取压法）, as shown in Fig. 4-10. Ring chamber tapping is the most common pressure taking methods. Under the premise of strict processing, manufacturing and installation quality, this pressure taking method can achieve higher measurement accuracy. When the length of the front and back straight pipe segments of the throttling device can meet the requirements, separate bore tapping can also be used for pressure taking. Note, however, that the distance between the farthest edge of the pressure-tapping hole and the front-end face of the throttling device should not exceed 0.03D. The bore diameter shall not exceed 0.03D, but not less than 4mm, and not more than 15mm (D is the inner diameter of the pipe).

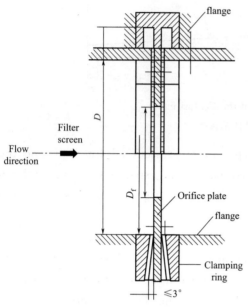

Fig. 4-10 Corner tapping

图 4-10 角接取压法

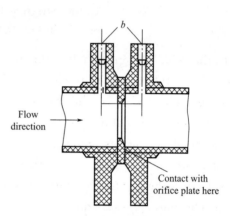

Fig. 4-11 Schematic diagram of flange tapping of standard orifice plate

图 4-11 法兰取压标准孔板示意图

The flange tapping (法兰取压法). The specific dimensions of the flange tapping: Both the upstream and downstream pressure taking centers are located at 25.4mm from the corresponding end faces on both sides of the orifice plate, as shown in Fig. 4-11. The flange tapping is convenient for processing and installation. At present, the application of the flange tapping in industry has been quite common.

The theoretical tapping (理论取压法): The center of the upstream pressure port is located at the front end of the pipe equal to the pipe diameter of the orifice plate, and the center of the downstream pressure port is located at the minimum cross section of the flow stream (namely, the vena contracta). In the derivation of the theoretical equation of the throttling device, the pressure difference taken out by these two sections is adopted, so it is called the theoretical tapping. The minimum cross-sectional area of the vena contracta always varies with the aperture ratio and flow rate, but the pressure port must be fixed at a certain position. So, the flow coefficient cannot be kept constant throughout the entire flow measurement range. In addition, because the pressure taking point is far away from the end surface of the orifice plate, it is difficult to achieve the ring chamber pressure taking, which will have a certain impact on the accurate pressure measurement. The advantage of the theoretical tapping is that the measured pressure difference is large.

The span tapping (径距取压法): The distance from the upstream pressure port to the front end of the orifice plate is the pipe diameter D, and the distance from the downstream pressure port to the front end of the orifice plate is $D/2$, so it is also called D-$D/2$ tapping. The differential pressure measured by the span tapping is smaller than that by the theoretical tapping.

c. Pitot tube (测速管/皮托管) The pitot tube is a device used to measure the local velocity along a streamline (用来测量导管中流体的点速度). The structure of the device is shown in Fig. 4-12.

Pitot tube consists of two concentric tubes bent at right angles. The nozzle of the outer tube closes, with the pressure ports around the outer tube wall. The ends of the outer and inner tubes respectively connect to the legs of a liquid

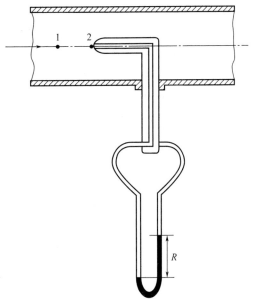

Fig. 4-12 Pitot tube
图 4-12 测速管

column pressure gauge. The opening of pitot tube is directly opposite the direction of the fluid flow in the pipe. During the measurement, the pitot tube is filled with the measured fluid. The velocity at point 1, located a short distance in front of the pitot tube nozzle, is u_1, and the static pressure is p_1. The velocity between point 1 and point 2 gradually slows down due to the obstruction of the impact opening of the pitot tube, pressure builds up. The flow velocity u_2 at opening point is zero (because the fluid in the pitot tube is not flowing), and the static pressure increases to p_2. The increase of hydrostatic head on the nozzle is due to the velocity head conversion of the fluid between points 1 and 2, so the hydrostatic head measured at point 2 is:

$$\frac{p_2}{\rho g} = \frac{p_1}{\rho g} + \frac{u_1^2}{2g} \tag{4-10}$$

Where, ρ is the density of the fluid, kg/m^3.

That is, the inner tube of the pitot tube is measured by the sum of the hydrostatic head and the dynamic head at the nozzle, which is collectively called the stamping head (在测速管的内管所测得的为管口所在位置的流体静压头之和, 合称为冲压头).

The outer wall of the pitot tube is parallel to the flow direction, there is no velocity component perpendicular to its opening. Therefore, the hydrostatic head $p_1/\rho g$ is measured on the pressure port on the outer wall of the pitot tube. Since the diameter of the pitot tube is very small, generally 5-6mm, the position height of the pressure port and the opening of the inner tube can be considered on the same horizontal line. The pressure head difference shown on the liquid column pressure gauge at the end of the pitot tube is the velocity head $u^2/2g$ on the horizontal line at nozzle:

$$\Delta h = \frac{p_2}{\rho g} - \frac{p_1}{\rho g} = \frac{p_1}{\rho g} + \frac{u_1^2}{2g} - \frac{p_1}{\rho g} = \frac{u_1^2}{2g} \tag{4-11}$$

or
$$u_1 = \sqrt{2g\Delta h} \tag{4-12}$$

Where, u_1 — the point velocity of the fluid on the horizontal line at nozzle, m/s;

h — the pressure head difference of the liquid pressure gauge, m;

g — gravity acceleration, $g = 9.81 \text{m/s}^2$.

If the opening of the pitot tube is aligned at the center line of the conduit, the measured point velocity is the maximum velocity of the fluid, written as Eq. (4-12), on the conduit section.

$$u_{max} = \sqrt{2g\Delta h} = \sqrt{\frac{2gR(\rho_i - \rho)}{\rho}} \tag{4-13}$$

Where, R — the reading on the liquid column pressure gauge, m;

ρ_i — density of indicating liquid, kg/m^3;

ρ — density of the fluid, kg/m^3.

Based on u_{max},

$$Re_{max} = \frac{du_{max}\rho}{\mu} \tag{4-14}$$

From the values of u/u_{max} found in Fig. 4-13, the average velocity $\bar{u}$ of the fluid on conduit cross section of can be obtained. Thus, the flow rate of the fluid in the conduit is

$$Q = AU = \frac{\pi}{4}d^2 u \tag{4-15}$$

Where, Q — flow rate of fluid, m^3/s;

A — the cross-sectional area of the conduit, m^2;

d — the inner diameter of the conduit, m.

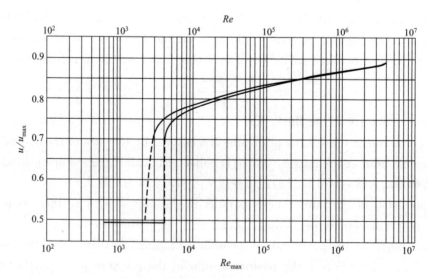

Fig. 4-13 u/u_{max} versus Re and Re_{max}

图 4-13 u/u_{max} 与 Re 和 Re_{max} 的关系

To improve the accuracy of the measurement, the pitot tube must be mounted on the straight pipe section and parallel to the axils of the straight pipe. Distance from nozzles to places where vortex is generated (e.g. elbows, transition pipe and valves, etc.) must be greater than 50 times the diameter of the straight pipe. Under such conditions, the velocity distribution of the fluid in the straight pipe is stable, and the point velocity determined at the centerline of the straight pipe is the maximum velocity. The pitot tube must be calibrated before use.

The structure of pitot tube is simple and has a small head loss for fluid. It is characterized by only measuring the point velocity and can be used to measure the fluid velocity distribution curve.

(3) Velocity-type flowmeter (速度式流量计)

① Turbine flowmeter (涡轮式流量计) Turbine flowmeter is a velocity flowmeter. It has high accuracy up to more than 0.5 level, rapid response, wide range and linear scale. Therefor, it is increasingly widely used in industrial production. The structure of the turbine flow transducer is shown in Fig. 4-14. The turbine is placed in a ball bearing with little friction, and a magnetoelectric device consisting of magnetic steel and induction coil is mounted on the housing of the magnetoelectric induction converter. When the fluid flows through the transducer, it will drive the turbine to rotate. The electromagnetic pulse signal is induced on the magnetoelectric induction converter, which is amplified and sent to the display instrument of a digital frequency integrator to measure the instantaneous value and accumulated value of the flow.

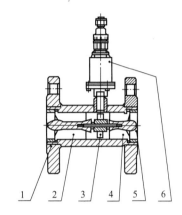

Fig. 4-14 Structure of turbine flow transducer

图 4-14 涡轮流量变送器

1—Shell units (壳体组件); 2—Forth leading bogie units (前导向架组件); 3—Impeller units (涡轮组件); 4—Real leading bogie units (后导向架组件); 5—Clamping ring (压紧圈); 6—Electromagnetic induction converter with amplifier (带放大器的磁电感应转换器)

As the fluid flows through the transducer, the rotating turbine makes the magnetic conductive blades periodically change the magnetic resistance value of the magnetic circuit in the detector, so that the magnetic flux through the induction coil changes accordingly. Thus, an electric pulse signal is generated at both ends of the induction coil. With a certain flow range, the frequency f of the electric pulse is proportional to the volume flow Q of the medium flowing through the transducer. In this way, the display instrument can calculate the instantaneous flow and accumulated flow over a certain time by the number of pulses. That is:

$$f = KQ \tag{4-16}$$

The proportionality constant K of the transducer are influenced by the viscosity changes of measured medium. When measuring a liquid flow with low viscosity, if the nominal diameter of turbine flow transducer $D_g > 25\text{mm}$, the water calibration results of the manufacturer can be directly used. Otherwise, to ensure sufficient accurate measurement results, the user should re-calibrate the instrument constant with the measured medium. In addition, constant K is almost only related to its geometric parameters except for the viscosity of the medium. Therefore, after the design and manufacture of a transducer is completed, its instrument constant has been determined, and this value can only be accurately obtained after calibration. Usually, the manufacturer calibrates the factory turbine transducer with clean water at normal temperature, and then gives the instrument constant on the check sheet.

The turbine flowmeter should be installed horizontally. The flow direction in the pipeline should be consist with the direction of arrow on the transducer label. The length of the straight pipe section before and after the inlet and outlet should not be less than $15D$ and $5D$, and the valve regulating the flow should be installed $5D$ away from the back straight pipe. To avoid the impurities in the fluids such as particles, fibers, ferromagnets, etc. from blocking the turbine blades and reducing bearing wear, a 20-60 mesh (0.25-0.85mm) filter should be installed in the front of the straight pipe before transducer. After a period of use, the filter should be removed and cleaned periodically according to the specific situation. The transducer should be installed unaffected by the external electromagnetic field, otherwise, a shield case should be added on the magnetic-electric induction converter of the transducer. Turbine flow transducer and the secondary display instrument should be well grounded, and the connecting cables shall be shielded cables.

Due to the high-speed rotation of the impeller during operation, wear occurs even with good lubrication. Thus, after a period of use, if the turbine transducer cannot work normally due to wear, the shaft or bearings should be replaced and cannot be used until re-calibrated.

② Electromagnetic flowmeter (电磁流量计)　Electromagnetic flowmeter is a kind of instrument applying the principle of induced potential generated by the movement of conductive fluid in magnetic field. According to the law of electromagnetic induction, when a conductor moves in a magnetic field and cuts the magnetic line, an induced potential will generate in the conductor. The induced potential is linear to the volume flow, so if an electrode is inserted on either side of the pipeline, the induced potential can be drawn out and the flow is indicated by the meter. All conductive liquid can be measured by electromagnetic flowmeter, which has a wide range of applications and can be used to measure the flow of a variety of corrosive acid, alkali, salt solutions and conductive liquids containing solid particles, such as mud or fiber. Since the electromagnetic flowmeter itself is easy to disinfect, it can also be used for the flow measurement of the pharmaceutical industry and food industry with special hygiene requirements, such as plasma, milk, fruit juice, alcohol, etc. In addition, it can also be used for flow measurements of large pipes of tap water and sewage.

(4) Volume flowmeter (容积式流量计)

① Elliptical gear flowmeter (椭圆齿轮流量计)　Elliptical gear flowmeter is a kind of positive displacement flowmeter, which is used to accurately measure the flow rate or instantaneous flow of liquid in the pipeline continuously or discontinuously. It is especially suitable for flow measurement of medium with high viscosity such as heavy oil, polyvinyl alcohol and resin.

② Wet gas flowmeter（湿式流量计） The instrument belongs to displacement flowmeter. It is a common instrument in the laboratory, mainly composed of a round drum shell, rotating drum and drive counting device, as shown in Fig. 4-15. The rotating drum is composed of a cylinder and four curved blades forming four small chambers of equal volume. The lower half of the drum submerges in water. The water amount is indicated by the water level device. Gas enters each chamber in turn from the intake pipe 9 in the middle of back, and expels from the top, forcing the drum rotation. The number of revolutions displays the volume by the counter and pointer on the dial by the counting mechanism. With a stopwatch, the gas flow can be measured directly.

As shown in Fig. 4-15, when the gas enters by the intake pipe, the B chamber is taking in, the C chamber is beginning to take in, and the D chamber will exhaust. A wet gas flowmeter can be used directly to measure the gas flow or as a standard instrument to calibrate other flowmeters.

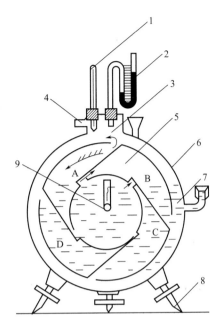

Fig. 4-15　Wet gas flowmeter
图 4-15　湿式气体流量计结构简图
1—Thermometer（温度计）；2—Manometer（压差计）；3—Horizon level（水平仪）；4—Exhaust pipe（排气管）；5—Drum（转鼓）；6—Shell（壳体）；7—Water level indicator（水位器）；8—Adjustable foot（可调支脚）；9—Intake pipe（进气管）

4.3　Temperature measurement（温度测量）

In chemical industry production and scientific experiments, temperature is the physical quantity representing the hot and cold degree of objects, and is often an important parameter for measurement and control. The temperature cannot be directly measured, but only indirectly by means of heat exchange between cold and hot objects, as well as some physical properties of objects that vary with degrees of cold and heat.

The temperature can be measured into two categories: contact and non-contact type.

The principle of contact type is that thermal equilibrium reaches within a long enough time after two objects contact. The temperature of the two unbalanced objects is equal, so that the measuring instrument can measure the temperature of the object.

Non-contact is the use of heat radiation principle. Sensitive element of the measuring instrument does not require contact with the measured material. It is commonly used for measurements where motion bodies and thermal capacity is small or particularly high temperatures. Various thermometers and operating principles are listed in Table 4-3.

Table 4-3 Classification and working principle of the thermometer
表 4-3 温度计的分类及工作原理

Classification of the thermometer			Working principle	Test range /℃	Main features
Contact thermometer	Expansion type （膨胀式）	Liquid-expansion type	Utilize the properties of liquids(mercury, alcohol) or solids (bimetallic sheets) that expand when heated	−200-700	Simple structure, low price, generally only used for on-site measurement
		Solid-expansion type			
	Pressure gauge type （压力表式）	Vapor-pressure type	Utilize the properties of a gas, liquid, or saturated vapor of certain liquids that change in volume or pressure when heated	0-300	Simple structure; explosion-proof; not afraid of vibration; used for short distance transmission; low accuracy; large hysteresis
		Liquid-pressure type			
		Steam type			
	Thermal resistance type （热电阻式）	Metal thermal resistance	Utilize the properties of changing the resistance of a conductor or semiconductor when heated	−200-850	High accuracy; long-distance transmission; suitable for low and medium temperature measurement; difficult to measure point temperature due to large volume
		Semiconductor thermistor		−100-300	
	Thermocouple type （热电偶式）		Utilize thermoelectric properties of objects	0-1600	Wide measurement range; long-distance transmission; suitable for medium and high temperature measurement, and need to be cold end temperature compensation; measurement accuracy is low in low temperature area
Non-contact thermometer	Optical type （光学式）		Utilize the properties that the radiant energy of a body varies with temperature	600-2000	Suitable for the cases where the temperature cannot be measured directly; mostly used for high temperature measurement; the measurement accuracy is affected by environmental conditions, and the error can be reduced only after the measurement value is corrected
	Colorimetric type （比色式）				
	Infrared type （红外式）				

(1) Expansion thermometer（膨胀式温度计）

The instrument used to measure the temperature according to the principle of liquid thermal expansion is called a liquid expansion thermometer, such as a glass tube thermometer. Using the nature of the length of solids as a function of temperature, the instrument for measuring temperature is called a solid expansion thermometer, such as a bimetal thermometer.

① Glass tube thermometer（玻璃管温度计）　The glass tube thermometer is made using the principle that temperature-measuring material（mercury, alcohol, toluene, kerosene, etc.）in the glass temperature sensing bubble expands in hot and contracts in cold ［玻璃管温度计是利用玻璃感温泡内的测温物质（水银、酒精、甲苯、煤油等）受热膨胀、遇冷收缩的原理进行温度测量的］. The glass tube thermometer is the most commonly used instrument for temperature measurement. It has the advantages of simple structure, cheap price, convenient reading and high precision. The measuring range is $-80\text{-}500℃$. Its disadvantage is that it is easy to damage and cannot make amend after damage. Mercury thermometers and organic liquids（such as ethanol）thermometers are the most commonly used in laboratories. Mercury thermometers have a wide measuring range, uniform scale and accurate readings, but damage can cause mercury pollution. The coloring reading of organic liquid（ethanol, benzene, etc.）thermometer is obvious, but because the expansion coefficient varies with temperature, the scale is uneven and the reading error is large.

Glass tube thermometer according to its purpose and use occasion can be divided into glass thermometer with metal protection tube, electric contact glass thermometer, standard mercury thermometer.

a. Glass thermometer with metal protective tube（带有金属保护管的玻璃温度计）　When glass tube thermometer is used in industrial process, to prevent the glass thermometer from being broken and to make the glass thermometer reliably fixed to the thermometer device, industrial glass thermometer is equipped with a metal protection tube. According to the shape of the internal standard glass thermometer, the glass thermometer with a metal protection tube has three forms: straight, 90° angular and 135° angular shape.

b. Electric contact glass thermometer（电接点玻璃温度计）　It uses mercury as a conductive dielectric to form a control circuit with an electronic relay, which is used to conduct over-limit alarm or off-on control of a certain temperature change. The working principle is: When the mercury rises to the contact with the temperature change, the control circuit is turned on, playing the role of limit alarm or control.

The electric contact glass thermometer is divided into adjustable or fixed forms according to whether the working contacts can be adjusted. The shape of the adjustable electric contact glass thermometer is shown in Fig. 4-16.

c. Standard mercury thermometer　The precision mercury thermometer used to calibrate the tested thermometer in the comparison method is called the standard mercury thermometer. Standard mercury thermometers are manufactured in complete sets, each with several thermometers. Temperature interval of each thermometer is small with zero marks. For example, a set of first-class standard mercury thermometer has 9 thermometers (0-100℃, the minimum division value is 0.05℃, the rest of the range is 0.1℃) and a set of 13 thermometers (the minimum division value is 0.05℃). A set of second-class standard mercury thermometer has 7 thermometers, the minimum division value is 0.1℃, it is a standard appliance commonly used in factories.

The glass pipe thermometer should be installed in a place convenient for reading on the equipment without large vibration and not susceptible to collision, especially for the organic liquid glass tube thermometer, which is easy to interrupt the liquid column if the vibration is very large. The center of the temperature bubble of the glass tube thermometer should be at the most sensitive place of temperature change (e.g. at the maximum flow rate in the pipe). Read position is at the highest point of the convex surface, and for the organic liquid glass tube thermometer at the lowest point of the concave surface. To measure the temperature accurately, when measuring the object temperature with a glass tube thermometer, the exposed liquid part of the thermometer should be corrected. An additional thermometer must be attached to the main thermometer, see Fig. 4-17. For example, in the measurement, the upper part of the mercury column is exposed outside the measured object, the temperature measured is not the temperature of the object to be tested, so it must be corrected as follows:

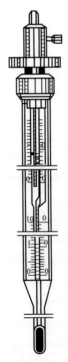

Fig. 4-16 Adjustable electric contact glass thermometer

图 4-16 可调式电接点玻璃温度计

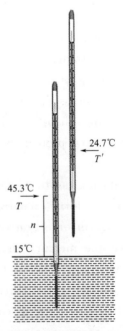

Fig. 4-17 Temperature correction for exposed liquid part

图 4-17 对露出液体部分的温度校正

$$\Delta T = \frac{n(T - T')}{6000} \tag{4-17}$$

Where, n—exposed part of the mercury column height (temperature scale);

T—temperature indicated by the thermometer;

T'—the intermediate temperature around the exposed part (measured with another thermometer);

1/6000—the difference of expansion coefficient between glass and mercury.

Then, the actual temperature after correction $= T + \Delta T$.

Calibration is required for accurate temperature measurements with glass tube thermometers. There are two methods of calibration: Comparison with standard thermometers under the same conditions using pure phase transition points such as ice-water and water-vapor systems. Insert the calibrated glass tube thermometer and the standard thermometer into the constant temperature tank. After the temperature of the constant temperature tank is stable, compare the indicated values of the tested thermometer and the standard thermometer. The thermometer can also be corrected with the phase transition temperature of ice-water, water-vapor.

a. Calibration for 0℃ with a mixture of water and ice Fill a 100mL beaker with crushed ice or ice cubes, then fill it with distilled water until the liquid level reaches 2cm below the ice surface. Insert a thermometer so that the scale is easy to observe or expose 0℃ on the ice surface. Stir and observe the change of the mercury column. When the indicated temperature is constant, record the reading. Be careful not to make the ice completely dissolved.

b. Calibration for 100℃ with water and steam Calibration thermometer is shown in Fig. 4-18 with the plug gap to balance the pressure inside and outside the test tube. Add a small amount of zeolite and 100mL of distilled water to the test tube. Adjust the thermometer so that the mercury bulb is 3cm above the liquid level. Heat it with low fire and condense the vapor on the tube wall to form a ring, control the fire to make it about 2 cm above the mercury ball. If a droplet is on the mercury sphere, there is thermal equilibrium between liquid and gas. When the temperature is constant, observe the mercury column reading and record the reading. After further pressure correction, it is a corrected 100℃.

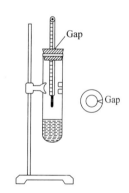

Fig. 4-18　Thermometer calibration device

图 4-18　温度计校正装置

② Bimetallic thermometer（双金属温度计） Bimetal thermometers usually use bimetal elements as temperature sensitive elements. Bimetallic elements are made by combining two metals with different linear expansion coefficients. When the measured temperature changes, the metal sheet is bent due to the different elongation produced by the two metal sheets, thus converting the temperature change to a shift change at the free end of the bimetal sheet. This bimetallic thermometer is stronger than the glass thermometer, and mercury-free. It applies in industrial measurement instead of the mercury thermometer due to a certain shock resistance and convenient reading. Its accuracy is generally lower than mercury thermometer.

(2) Pressure thermometer (压力式温度计)

A pressure thermometer is made from the principle of the temperature-dependent pressure of a gas or a saturated vapor of a certain liquid filled in a closed container (利用封闭在密闭容器中填充气体或某种液体的饱和蒸气的压力随温度变化的原理制成的温度计称为压力式温度计). It can be divided into gas pressure thermometer, vapor pressure thermometer and liquid pressure thermometer according to the different filling substances.

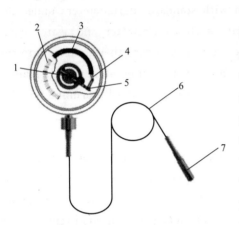

Fig. 4-19 Operating principle of a pressure thermometer

图 4-19 压力式温度计的作用原理

1—Pointer (指针); 2—Dial (刻度盘); 3—Spring tube (弹簧管); 4—Linkage (连杆); 5—Transmission mechanism (传动机构); 6—Capillary (毛细管); 7—Temperature wrap (温包)

Fig. 4-19 shows a pressure thermometer, which is a closed system, composed of temperature wrap, capillary and spring tube. The system is filled with temperature sensitive substances, such as nitrogen, mercury, xylene, toluene, glycerin and low boiling point liquids, such as chloromethane, chloroethane and so on. During measurement, the temperature wrap is placed in the measured medium. When the temperature of the measured medium changes, the pressure of the temperature-sensitive substance in the temperature wrap changes due to heat. Pressure increases as temperature increases and decreases as temperature decreases. The change of pressure is transferred to the spring tube through the capillary, one end of the spring tube is fixed, and the other end is free. The displacement caused by the pressure change, through the transmission mechanism, drives the pointer to indicate the corresponding temperature change value.

The temperature wrap is directly in contact with the measured medium to feel its temperature change. Therefore, the temperature wrap is required to have high mechanical strength, small coefficient of expansion, high thermal conductivity and corrosion resistance. The temperature wrap is commonly made of purple copper pipe, seamless steel pipe or stainless-steel pipe, with an outer diameter of 12-22mm and a length of 65-435mm. One end of the pipe is welded with a cover plate, and the other end is connected to the capillary by a short tube of 235-300mm. The short tube is provided with a fixed screw thread for mounting the temperature wrap.

The capillary is used as a conduit for connection and transfer pressure between the temperature wrap and the spring tube pressure gauge, generally made of copper or stainless steel cold drawn seamless pipe with an internal diameter of 0.15-0.5mm, length of 20-60m. Because capillary tubes are easily damaged, they are usually protected by metal hoses or wrappers braided with copper or galvanized steel wires.

The features of pressure thermometer（压力式温度计的特点）are：

① The maximum capillary length of pressure thermometer can be up to 60m, so the thermometer can be measured in situ, but also can be displayed, recorded, alarm and adjusted to the measured temperature at a long distance of 60m.

② The structure of the pressure thermometer is simple, low price and clear scale, suitable for temperature measurement of gas, vapor or liquid in industrial equipment within $-80\text{-}500\,^\circ\!\mathrm{C}$. The maximum pressure of the tested medium is 6MPa.

③ Except for the electric contact pressure thermometer, other forms of thermometer do not have power supply, and there will be no spark in use. Therefore, it has explosion-proof performance, suitable for the temperature measurement in flammable and explosive environment.

④ The values of the pressure thermometer are transmitted through capillaries with a long lag time, that is, the time constant is large. In addition, the mechanical strength of capillary is poor, so the capillary is easy to damage, and not easy to repair after damage.

The appropriate range of thermometer shall be selected according to the actual temperature measured. Should not exceed its allowable temperature measurement range in use, to avoid aging and affecting the service life. To calibrate before installation, a simple way is to reference check its room temperature display value with a standard glass mercury thermometer, then check its indicator standard at a certain point in hot or boiling water. During the calibration process, pay attention to observe whether the transmission system is flexible and whether the pointer moves smoothly. When qualified, the thermometer can be installed for temperature measurement. During use, keep the meter clean to facilitate reading. At the same time, attention should be paid to maintenance, do not make the thermometer temperature sense part rot and rust.

（3）Thermal resistance thermometer（热电阻温度计）

Thermal resistor is a widely used temperature sensing element for temperature measurement in industry, and has the advantages of simple structure, high accuracy and convenient use. Thermal resistor, used with the secondary instrument, can remotely transmit, display, record and control the temperature of liquid, gas, vapor and solid surfaces in the temperature range of $-200\text{-}600\,^\circ\!\mathrm{C}$.

The temperature measurement principle of thermal resistance is based on the resistance value of the metal or semiconductor changing with the temperature, and then the resistance value of the thermal resistance is measured by the display instrument to obtain the temperature value corresponding to the resistance value. A temperature measuring device composed of thermal resistance, connecting wire and display instrument is called a thermal resistance thermometer.

Structure of the ordinary thermal resistance（普通热电阻的构造）is shown in Fig. 4-20, the ordinary thermal resistance is mainly composed of resistor, lead wire, insulator, protective casing and junction box. Armored thermal resistance（铠装热电阻）developed in recent

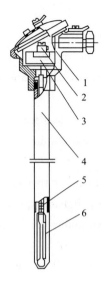

Fig. 4-20 Structure of thermal resistance

图 4-20 热电阻的构造

1—Terminal box（接线盒）；
2—Terminal column（接线柱）；
3—Terminal seat（接线座）；
4—Protective casing（保护套管）；
5—Insulator（绝缘子）；
6—Resistor（电阻体）

years. It is composed of metal protection sleeve, insulation material and resistor. It has the following characteristics:

① Small inertia and quick response. For example, an ordinary platinum resistor with a protective casing diameter of 12mm has a time constant of 25s; while the time constant of armored thermal resistance with metal sleeve diameter of 4.0mm is about 5 seconds.

② Bendability. Armored thermal resistance, except the head, can bend in any direction, so it is suitable for the temperature measurement of small equipment with more complex structure, and has good vibration resistance and impact resistance performance.

③ Due to the cover of magnesium oxide insulation material and metal casing protection, the thermoelectric wire is not easily eroded by harmful media, so its life is longer than the general thermal resistance.

At present, the thermal resistance produced by industrial standardization is mainly platinum resistance, copper resistance and nickel resistance. Platinum and copper resistors are widely used in industry.

① Platinum resistor（铂电阻） Platinum is an ideal material for manufacturing thermoelectric resistor, which is easy to purify, has high stability and good replicability in oxidative medium. The relationship between resistance and temperature is approximately linear and has high measuring accuracy. However, at high temperatures, platinum is susceptible to reductive media damage, and the texture becomes brittle. In the temperature range of 0-850℃, the relationship between platinum resistance and temperature can be expressed as follows:

$$R_t = R_o(1 + At + Bt^2) \tag{4-18}$$

Where, A, B—constants, obtained by experiments; for common industrial platinum resistors: $A = 3.90802 \times 10^{-3}$; $B = 5.80195 \times 10^{-7}$;

t—temperature, ℃;

R_o—resistance value at 0℃;

R_t—resistance value at temperature t, ℃.

② Copper resistor（铜电阻） Copper resistance is characterized by a linear relationship between resistance and temperature with a relatively large resistance temperature coefficient. The material is easy to purify and relatively cheap. However, its resistivity is low, and easy to oxidize. Copper resistors can be used when there is no special restriction.

In the range of －50-150℃, the copper resistance-temperature relationship is:

$$R_t = R_o(1+\alpha t) \tag{4-19}$$

Where, α—copper resistance temperature coefficient, $\alpha = (4.25\text{-}4.28)\times 10^{-3}\,℃^{-1}$.

The index number of industrial copper resistance in China is Cu50 ($R_o = 50\,\Omega$), Cu100 ($R_o = 100\,\Omega$).

③ Semiconductor resistance (thermistor) [半导体电阻（热敏电阻）]　Thermal resistance made from semiconductor materials is called thermistor. Most semiconductor thermistors have negative resistance temperature coefficient, its resistance value decreases with the increase of temperature. Although the irregular motion of temperature raising particles causes a slight decrease in free electron mobility, the number of free electrons increases faster with temperature, so the temperature increases its resistance value.

Thermistors are available in a variety of shapes. The thermistors used as thermometers are made of small spherical thermistors wrapped in glass or other films. The standard of the spherical thermistor body is a small ball with a diameter of 1mm to 2mm and sealed with two 0.1mm platinum wires as conductors, as shown in Fig. 4-21.

Semiconductor thermistors are usually made of oxides of some metals such as iron, nickel, molybdenum, titanium and copper, can measure the temperature of 100-300℃. It has many advantages: High resistance temperature coefficient (about 3%-6%), high sensitivity; high resistivity, so small volume; the resistance value is very large, so the influence of the

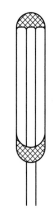

Fig. 4-21　Spherical thermistor
图 4-21　球状热敏电阻

resistance change of the connecting wire can be ignored; simple structure; thermal inertia is low.

(4) Thermocouple thermometer (热电偶温度计)

Thermal resistance transforms a temperature signal into a potential (mV) signal with a millivolt meter or transmitter enabling measurement of temperature or conversion of the temperature signal. It has the advantages of stable performance, good reproducibility, small volume, and small response time.

① Measurement principle of thermocouple thermometer (热电偶的测温原理)　Two different conductors (or semiconductors) A and B form a closed circuit as shown in Fig. 4-22. When the connection temperature at the left and right ends of A and B are different, a potential, called thermal potential, generates in the circuit. The closed loop in Fig. 4-22 is called a thermocouple. Conductors A and B are called the thermal electrodes. The junction placed in the measured medium (the temperature is T) is known as the working side or the hot end; while the end with reference temperature T_0 is called the cold end.

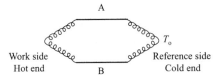

Fig. 4-22　Thermocouple closing circuit
图 4-22　热电偶闭合回路

In the thermocouple circuit, the thermoelectric potential E is associated with the temperature T and T_0 at both ends of the thermocouple. If T_0 is kept constant, the thermoelectric potential E depends only on T. In other words, for a certain thermocouple material, its thermoelectric potential E is only a function of the measured temperature T. After E is measured with a moving coil meter or potentiometer, the temperature to be measured is known.

② Common types of the thermocouple （常用热电偶的种类）

a. T-class: Copper （＋） vs Constantan （－） Such thermocouples are environmentally adaptable and corrosion resistant and can be used in vacuum, oxidized, reduced or inert gases. It is suitable for temperature measurements below 0℃ with a temperature limit of 371℃. In addition, this kind of thermocouples are cheap, can produce a large thermal potential. However, due to poor repeatability, the armored copper-constantan thermocouples are not available in the market.

b. J-class （Domestic TK class）: Iron （＋） vs Constantan （－） This type of thermocouple can be used in vacuum, oxidation, reduction or inert gases at temperatures below 760℃, but not in sulfur containing gases above 538℃. It is low-cost and the resulting thermal potential is large.

c. K-class （Domestic EU class）: Nickel and chromium alloy （＋） containing 10％ chromium versus nickel aluminum or nickel silicon alloy （－） containing 5％ nickel These thermocouples have better antioxidant properties than other metal thermocouples and are suitable for continuous use in oxidized or inert gas below 160℃ but not in reduced gas （unless a protective casing is added）. When used in sulfur-containing gas, a protective casing is required because sulfur damages the thermal electrode and causes intergranular corrosion, which will rapidly cause embrittlement and fracture of the negative thermal electrode wire. The thermoelectric performance is relatively consistent among such thermocouples, with large thermal potential, good linearity, wide temperature measurement range and cheap price, suitable for acidic environment. It is the most commonly used thermocouple in industrial production. The disadvantage is that in long-term use, the measurement accuracy will be affected by the change of thermoelectric properties due to the oxidation and deterioration of nickel and aluminum.

d. E-class Nickel chromium containing 10％ chromium （＋） versus Constantan （－） This type of thermocouple is suitable for use in oxidized or inert gases in the temperature range of －250-871℃. When used in reducing or alternating oxidizing and reducing environments, the limitations are the same as those of K-class. In all kinds of commonly used thermocouples, class E thermocouples produce high electric potential per degree. $dE_{AB}(t, t_o)/dt = 74.5 \mu V/℃$ at 200℃, so such type of thermocouple is widely used.

e. R, S-class （S-class is equivalent to domestic LB-class）: R is the platinum rhodium alloy containing 13％ platinum （＋） versus platinum （－）, S is the platinum rhodium alloy containing 10％ platinum rhodium （＋） versus platinum （－）. This kind of thermocouple is high temperature resistant, suitable for continuous use in oxidation or inert gas below

1399℃ (domestic 1300℃). At high temperature, it is vulnerable to reducing vapor and metal vapor, leading to changes in thermocouple characteristics, so it is not suitable for use in reducing gas. Due to the easy access to high purity platinum and platinum rhodium, this kind of thermocouples have high replication accuracy and stable performance, and can be used for the measurement of accurate temperature or as a standard thermocouple. The disadvantages are weaker thermoelectric potential, nonlinear thermoelectric properties, and higher cost of precious metal materials.

f. B-class (domestic LL-class): platinum-rhodium alloy with 30% rhodium (+) versus platinum-rhodium alloy with 6% rhodium (−)　This thermocouple can be used for long-term measurement of high temperatures up to 1600℃. It has stable performance and high accuracy, suitable for use in oxidative and neutral medium, and not in reductive gas. Its disadvantage is that the thermoelectric potential generated is small, and expensive.

g. Domestic EA-class: nickel-chrome alloy containing 10% chromium (+) versus Kao copper (Ni-Cu alloy containing 44% nickel) (−)　Such thermocouples are suitable for use in reductive or neutral media. If long-term use, temperature should not exceed 600℃. It is characterized by high thermoelectric sensitivity, large thermoelectric potential and low price. However, the temperature range is low and narrow, and the copper alloy wire is easy to oxidize and deteriorate.

③ Temperature compensation at the cold end of the thermocouple (热电偶冷端的温度补偿)　According to the principle of thermocouple temperature measurement, the thermoelectric potential is a single value function of measured temperature only when the temperature of the cold end of the thermocouple remains constant. In application, because the thermocouple's working end (hot end) is very close to the cold end, and the cold end is exposed to space, it is easy to be affected by ambient temperature fluctuations, so the cold end temperature is difficult to keep constant. Therefore, the following treatment methods are adopted.

a. Compensation wire method (补偿导线法)　To keep the cold end temperature of the thermocouple constant (preferably 0℃), the thermocouple can certainly be made very long so that the cold end is away from the working end, and placed together with the measuring instrument at constant temperature or in a place with low temperature fluctuations (e.g. centralized in the control room), but this method makes the installation inconvenient and costs many expensive metal materials. Therefore, the cold end of the thermocouple is generally extended by a wire (called compensation wire), which has the same thermocouple as the connected thermocouple within a temperature range (0-100℃), and its material is cheap metal. for commonly used thermocouples, such as platinum-rhodium-platinum thermocouples, the compensation wire is copper-nickel-copper; for nickel-chromium-nickel-silicon thermocouples, copper-constantan compensation wires are used; for thermocouples made of cheap metal such as nickel-chromium-Kao copper, iron-Kao copper and copper-constantan, their own materials can be used to extend the cold end to where the temperature is constant.

It must be noted that the application of compensation wires is only meaningful when the newly shifted cold end temperature is constant or when the dispensing instrument itself has an automatic cold end temperature compensation device. If the newly moved cold end is still in a high temperature or fluctuation place, then the compensation wire at this time has completely lost its due role. Therefore, the cold end of the thermocouple must be properly positioned.

In addition, the temperature of the thermocouple and compensation wire connection should not exceed 100℃, otherwise it will introduce new errors due to different thermoelectric characteristics.

b. The correction of cold-end temperature method（冷端温度校正法） The temperature-thermal potential relationship curve (scale characteristic) of the thermocouple is obtained when the cold end temperature remains 0℃, and the instrument used with it is calibrated on the basis of this curve. Therefore, although compensation wires have been used to extend the cold end of the thermocouple to a constant temperature, the instrument indication must be corrected as long as the cold end temperature is not equal to 0℃.

For example, if the cold end temperature is higher than 0℃ but constant at t_0℃, the measured thermoelectric potential is less than the indexing value of the thermocouple. At this point, in order to obtain the real temperature, the following formula can be used for correction:

$$E(T,0)=E(T,t_0)+E(t_0,0) \tag{4-20}$$

c. Ice bath method（冰浴法） To avoid frequent correction, ice bath method can be used to keep the cold end temperature constant at 0℃. Under laboratory conditions, the cold end is usually placed in a test tube with insulating oil and then placed in a container filled with a mixture of ice water to keep the cold end at 0℃.

d. Compensation bridge method（补偿电桥法） As shown in Fig. 4-23(a), a resistance with a large temperature coefficient (usually copper resistance) is connected in the circuit, then the total potential in the thermocouple circuit is $E = E_0 + IR_{Cu} = E_0 + IR_0(1+\alpha t)$. Where E is the potential of the thermocouple, α is the resistance temperature coefficient of the copper resistance; R_0 is the resistance value of copper resistance at 0℃. When the cold end of the thermocouple experiences the same temperature as the copper resistor, the voltage change on the copper resistor can compensate the change of thermoelectric potential caused by the change of the cold end temperature of the thermocouple. In practical applications, compensation bridges with low internal resistance are mostly used [as shown in Fig. 4-23(b)]. When the bridge is in balance, the bridge has no effect on the meter reading. Because the thermoelectric potential of a thermocouple is nonlinear relative to temperature, the compensation circuit composed of a copper resistor has a large error in a large temperature range. In this case, it is better to use the two copper resistance compensation methods as shown in Fig. 4-23(c). Please refer to relevant books for more information.

④ Display instrument（显示仪表） Thermocouple display instruments generally include moving-coil instrument（动圈式仪表）, DC potentiometer（直流电位差计）,

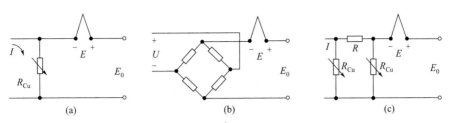

Fig. 4-23　Compensation bridge
图 4-23　补偿电桥

electronic potentiometer（电子电位差计），digital voltmeter（数字电压表） and so on. Among them，potentiometer is used more in the laboratory. The potential difference principle is measured based on the voltage balance method（or the voltage offset method），so the potentiometer is also known as the "voltage balance". That is，the known voltage is used to counterbalance the measured electric potential. When the current is equal to zero in the measurement loop，the known voltage value displayed at this time is the potential value measured.

For further information on how potential difference meters work，please consult relevant books and materials.

Appendix
附　录

Appendix 1　SXK-2 type high precision flow integrator
附录 1　SXK-2 型高精度流量积算仪

SXK-2 type high precision flow integrator, designed and supervised by our university, is shown in Fig. A1-1. Using modern electronic technology of integrated circuit, the product has high performance, easy to use and intuitive characteristics. It can be used with various turbine and various types of pulse transmitter. Display mode can be directly converted into of instantaneous flow with four digits (unit m^3/h or t/h) and cumulative flow with six digits (unit m^3 or t, to 0.2-order accuracy), can also display the line speed and cumulative total length number. It has the function of self-calibration and signal measurement with advanced indexes, performance and precision, and widely used in universities, scientific research institutes, petroleum, chemical, textile, metallurgy, and other industrial departments.

Fig. A1-1　Appearance of SXK-2type high precision flow integrator
图附 1-1　SXK-2 型高精度流量积算仪——外观

A1.1 Schematic diagram of measuring principle（测量原理示意图）(Fig. A1-2)

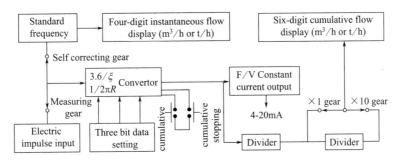

Fig. A1-2　Schematic diagram of measurement principle
图附1-2　SXK-2型高精度流量积算仪——原理示意图

A1.2 Use instructions（使用说明）

(1) Switch to measuring gear

(2) According to the factory calibration of turbine flux transmitter：ξ = pulse number/V, and the unit of V is dm^3.

(3) The instrument converts a pulse number into an internationally recognized unit of measurement m^3/h or t/h, so the instrument setting constant：$1/\xi$ (L/s) = (0.001m^3 × 3600h^{-1}) (1/ξ) = (3.6/ξ) m^3/h. In the following example, according to the factory calibration of the ξ into 3.6/ξ, the full relationship between each setting constant and the decimal place and unit of instantaneous and cumulative flow can be obtained.

Lw caliber D_g/mm	Flow range /m^3	ξ = pulse number/dm^3	3.6/ξ data setting (three-digit)	Instantaneous flow	Cumulative flow	
					×10gear	×1gear
10	0.2-1.2	3080.0-3090.0	0.001168-0.001165	0.01m^3/h	10dm^3	1dm^3
25	1.2-10	275.00-383.00	0.01309-0.01272	0.1m^3/h	0.1m^3	10dm^3
40	3-20	76.00-77.00	0.04736-0.04675	0.1m^3/h	0.1m^3	10dm^3
50	6-40	37.00-38.00	0.09769-0.09475	0.1m^3/h	0.1m^3	10dm^3
80	16-100	13.00-14.00	0.2769-0.2571	1m^3/h	1m^3	0.1m^3
100	25-160	8.00-9.00	0.4500-0.4000	1m^3/h	1m^3	0.1m^3
150	50-300	2.50-2.70	1.440-1.333	10m^3/h	10m^3	1m^3

(4) Regulation grading method for constant flow output 0-10mA or 4-20mA corresponding to flow：

① Turn the switch gear to self-correcting gear, set the dial "000", display all "0" means the flow of all "0". Adjust the 12 Ω variable resistance in the plate so that the constant current output 4mA, and then on the full flow, such as 40m^3/h, set the "400" to about 20mA, and the 200 Ω variable resistance in the plate is adjusted to 20mA. After several repeated adjustments, when dialing "000", the output is 4 mA and when dialing

"400", the output is 20mA. That is, 4-20mA corresponding to 0-40.0 m³/h. Dividing line error $<\pm 0.5\%$.

② According to the above adjustment method, it can be adjusted randomly as required.

③ Adjust the scale division, turn the switch to the measuring gear and set the data of $3.6/\xi$, then measure the instantaneous, cumulative flow and constant current output of 4-20mA.

(5) When the instrument is used with various turbine and various types of pulse transmitter, the setting constant of the instrument is:

① Measurement flow: $3.6/\xi$, showing instantaneous flow m³/h, cumulative m³.

② Measurement of linear velocity: $L/2\pi R$, displaying linear velocity m/s, cumulative total length m.

(6) The external wiring is at the back of the instrument, in strict accordance with the requirements of the external wiring marked on the schematic diagram (Fig. A1-3) (外接线在仪器后面,严格按照示意图所标出的要求外接线).

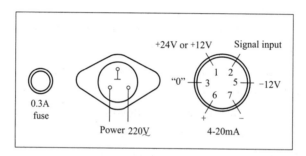

Fig. A1-3　SXK-2type high precision flow integrator——Schematic diagram of external wiring
图附 1-3　SXK-2 型高精度流量积算仪——外接线示意图

Appendix 2　ZW5433 three-phase digital meter
附录 2　ZW5433 三相数字电量表

The ZW5433 three-phase digital power meter is shown in Fig. A2-1. The test object of the digital meter is 45-65Hz power-frequency AC signal, which can be used to measure three-phase voltage, three-phase current, total active power, total reactive power, total power factor, and the frequency of A-phase voltage signal. The intuitive digital display of the instrument can effectively avoid reading errors. The wiring is simple, with three-phase three-wire, three-phase four-wire measurement mode, voltage and current rate can be set through the panel button on site. Not only for the measurement of true effective value, but also for the distortion waveform with a high measurement accuracy. The measurement data can be transmitted long distance through RS458 serial communication unit, over upper and lower limit alarm output can be achieved by relay. Panel mounted can directly replace the pointer meter.

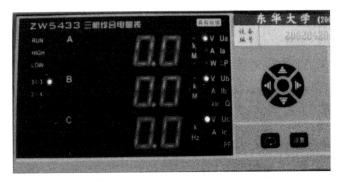

Fig. A2-1　ZW5433 three-phase digital power meter—Voltage measurement
图附 2-1　ZW5433 三相数字电量表——测量电压

A2.1　Definition of panel symbol（操作面板符号定义）

① Unit symbol（单位符号）

k：thousand；M：megabytes；V：volt；A：Ampere；W：Watt；Hz：Hertz.

② Key symbol（按键符号）

-up　-down　-left　-right　-alter　-set

A2.2　Power measurement（功率测定）

The "V" lamp of the three-phase digital power meter represents the measured voltage, the "A" lamp represents the measured current, and the "W" lamp represents the measured electric power. Under the interface shown in the above figure (the "V" lamp is the voltage measurement interface, but also the three-phase digital power meter start interface), press the "alter" button twice to switch to the power measurement interface shown below ("W" light is on), press the "alter" button twice again to switch to the original voltage measurement interface.

On the power measurement interface, if the "k" lamp is on, the power unit is kilowatts. If the "k" lamp is off, the power unit is watts. The power shown in the figure below (Fig. A2-2) is recorded as 1.542kW.

Fig. A2-2　ZW5433 three-phase digital power meter—Power measurement
图附 2-2　ZW5433 三相数字电量表——功率测量

Appendix 3　Instructions for hydrostatic balance
附录3　液体比重天平使用说明

The specific gravity of liquid samples at the top and bottom of the experimental distillation tower was measured by PZ-A-5 hydrostatic balance（液体比重天平）. Then, the mass percentage of the sample of ethanol solution according to the specific gravity value of the sample is determined referring to the specific gravity table of ethanol solution in Appendix 3.1（根据附录3.1乙醇溶液相对密度表，由测得样品的相对密度值查得相应的乙醇溶液样品的质量分数）. The principle and methods of liquid gravity balance are described as follows.

The hydrostatic balance has a measuring bob with a standard volume（$5cm^3$）and weight, which is immersed in the liquid to gain buoyancy, thus unbalance the beam. Then the riding code of the corresponding weight is placed in the V-shaped groove of the beam to restore the balance, so that the liquid specific gravity can be quickly measured, as shown in Fig. A3-1.

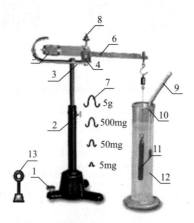

Fig. A3-1　Hydrostatic balance
图附3-1　液体比重天平

1—Horizontal adjusting screw（水平调节螺钉）; 2—Pillar fastening screw（支柱紧固螺钉）; 3—Bracket（托架）; 4—Agate blade seat（玛瑙刀座）; 5—Balance regulator（平衡调节器）; 6—Beam（横梁）; 7—Riding code (four)［骑码（4只）］; 8—Barycenter regulator（重心调节器）; 9—Thermometer（温度计）; 10—Thermometer holder（温度计夹）; 11—Measuring bob（测锤）; 12—Glass cylinder（玻璃筒）; 13—Equal weight（等重砝码）

Wash the measuring bob 11 and glass cylinder with pure water or alcohol and dry or wipe. Then loosen the pillar fastening screw 2. After lifting the bracket 3 to an appropriate height, tighten the pillar fastening screw 2, and place the beam 6 on the agate blade seat 4 of the bracket.

A3.1　Calibration of hydrostatic balance（液体比重天平的校正）

(1) Correct zero of the hydrostatic balance with water（用水校正液体比重天平的零点）

Fill the water in the measuring cylinder, then soak the measuring bob in the center of the water, and hang the other end on the small hook at the right end of the beam. Measure the water temperature in the measuring cylinder with a thermometer. Find out the density of the corresponding water by the density table of Appendix 3.2. Then place the riding code of corresponding weight based on the density of water in the V-shaped scale groove of the beam. Adjust the horizontal adjusting screw 1 to make the beam horizontal with the bracket fingertip to show balance. If balance cannot be adjusted, turn balance regulator 5 slightly until balance. Do not move the hydrostatic balance after zero correction. Empty the water in the cylinder, dry the measuring bob and cylinder for later use.

(2) Calibrate zero of hydrostatic balance with equal weight（用等重砝码校正液体比重天平的零点）

Hang the equal weight 13 to the right end of the beam. Adjust the horizontal adjustment screw 1 to balance the beam horizontally with the bracket tip. If balance is uncontrollable, turn balance regulator 5 slightly until balance. Thus, the zero point of the hydrostatic balance is adjusted.

A3.2 Measurement of specific gravity of liquids（液体相对密度的测量）

Pour the tested liquid into the glass measuring cylinder. Immerse the measuring bob in the center of the liquid. The beam is off balance due to buoyancy. Place the corresponding weight in the V-groove of the beam to balance the beam（horizontal to the tip of the bracket pin）. The sum of the riding code on the beam is the specific gravity value of the measured liquid. The reading method can refer to the table below.

The weight of the weights placed on the hook and in the V-shaped slot(放在小钩上与V形槽砝码重)	5g	500mg	50mg	5mg
The number represented by the 9th digit in the V-shaped slot(V形槽上第9位代表数)	0.9	0.09	0.009	0.0009
The number represented by the 8th digit in the V-shaped slot(V形槽上第8位代表数)	0.8	0.08	0.008	0.0008
The number represented by the 7th digit in the V-shaped slot(V形槽上第7位代表数)	0.7	0.07	0.007	0.0007
...	...	...	...	...

For example, the added riding codes of 5g, 500mg, 50mg and 5mg are the 9th, 6th, 2nd and 4th position respectively in the V-shaped scale slot of the beam, the specific gravity of the liquid can be read as 0.9624（例如，所加骑码5g，500mg，50mg，5mg，在横梁V形刻度槽位置分别为第9位，第6位，第2位，第4位，即可读出测量液体的比重为0.9624）. The reading is done by reading the V-groove scale in descending order of riding codes, i.e., the specific gravity value.

Appendix 3.1 Specific gravity table of ethanol solution
附录 3.1 乙醇溶液相对密度表

Mass fraction of ethanol/%	10℃	15℃	20℃	25℃	30℃	35℃	40℃
0	0.99973	0.99913	0.99823	0.99708	0.99568	0.99406	0.99225
1	785	725	635	520	379	217	034
2	602	542	453	336	194	031	0.98846
3	426	365	275	157	014	0.98849	663
4	258	195	103	0.98984	0.98839	627	485
5	098	032	0.98938	817	670	501	311
6	0.98946	0.98877	708	656	507	335	142
7	801	729	627	500	347	172	0.97975
8	660	584	478	346	189	009	808
9	524	442	331	193	031	0.97846	641
10	393	304	187	043	0.97875	685	475
11	267	171	047	0.97897	723	573	312
12	145	041	0.97910	753	573	371	150
13	026	0.97914	775	611	424	216	0.96969
14	0.97911	790	643	427	278	063	829
15	800	669	514	334	133	0.96911	670
16	692	552	387	199	0.96990	760	512
17	583	433	259	062	844	607	352
18	473	313	129	0.96997	697	452	189
19	363	191	0.96997	782	547	294	023
20	252	068	864	639	395	134	0.95856
21	139	0.96944	729	495	242	0.95973	687
22	024	818	592	348	087	809	516
23	0.96907	689	453	199	0.95929	634	343
24	787	558	312	048	769	476	168
25	665	424	168	0.95895	607	306	0.94991
26	539	287	020	738	442	133	810
27	406	144	0.95867	576	272	0.94955	625
28	268	0.95996	710	410	098	774	438
29	125	844	548	241	0.94922	590	248
30	0.95977	686	382	067	741	403	055

(continued)

Mass fraction of ethanol/%	10℃	15℃	20℃	25℃	30℃	35℃	40℃
31	823	524	212	0.94890	557	214	0.93860
32	665	357	038	709	370	021	662
33	502	136	0.94860	525	180	0.93825	461
34	334	011	679	337	0.93986	626	257
35	162	0.94832	494	146	790	425	051
36	0.94986	650	306	0.93952	591	221	0.92843
37	805	464	114	756	390	016	634
38	620	273	0.93919	556	186	0.92808	422
39	431	079	720	353	0.92979	597	208
40	238	0.93882	518	148	770	385	0.91992
41	042	682	314	0.92940	558	170	774
42	0.93842	478	107	729	344	0.91952	554
43	639	271	0.92897	516	128	733	332
44	435	062	685	301	0.91910	513	106
45	226	0.92852	472	085	692	291	0.90884
46	017	640	257	0.91868	472	069	660
47	0.92806	426	041	649	250	0.90845	434
48	593	211	0.91823	429	028	621	207
49	379	0.91995	604	208	0.90805	396	0.89979
50	0.92126	0.91776	0.91348	0.90985	0.90580	0.90168	0.89750
51	0.91943	555	160	760	353	0.89940	519
52	723	333	0.90936	534	125	710	288
53	502	110	711	307	0.89896	479	056
54	279	0.90885	485	079	667	248	0.88823
55	055	659	258	0.89850	437	016	589
56	0.90831	433	031	621	206	0.88784	356
57	607	207	0.89803	392	0.88975	552	122
58	381	0.89980	574	162	744	319	0.87888
59	154	752	344	0.88931	512	085	653
60	0.89927	523	113	699	278	0.87851	417
61	698	293	0.88882	446	044	615	180
62	468	062	650	233	0.87809	379	0.86943
63	237	0.88830	417	0.87998	574	142	705

(continued)

Mass fraction of ethanol/%	10℃	15℃	20℃	25℃	30℃	35℃	40℃
64	086	597	183	763	337	0.86905	466
65	0.88774	364	0.87948	527	100	667	227
66	541	130	713	291	0.86863	429	0.85987
68	374	0.87660	241	0.86817	387	0.85950	407
69	0.87839	424	004	579	148	710	266
70	602	167	0.86766	340	0.85908	470	025
71	365	0.86949	527	100	667	228	0.84783
72	127	710	287	0.85859	426	0.84986	540
73	0.86888	470	047	618	184	734	297
74	648	229	0.85806	376	0.84941	500	053
75	408	0.85988	564	134	698	257	0.83809
76	168	747	322	0.84891	455	013	564
77	0.85927	505	079	647	211	0.83768	319
78	685	262	0.84835	403	0.83966	523	074
79	442	018	590	158	720	277	0.82827
80	197	0.84772	344	0.83911	473	029	578
81	0.84950	525	096	664	224	0.82780	329
82	702	277	0.83848	415	0.82974	530	079
83	453	028	599	164	724	279	0.81828
84	203	0.83777	348	0.82913	473	027	576
85	0.83951	525	095	660	220	0.81774	322
86	697	271	0.82840	405	0.81965	519	067
87	441	014	583	148	708	262	0.80811
88	181	0.82754	323	0.81888	448	003	352
89	0.82919	492	062	626	186	0.80922	291
90	654	227	0.81797	362	0.80922	478	028
91	386	0.81797	529	094	655	211	0.79761
92	114	688	257	0.80823	384	0.79941	491
93	0.81839	413	0.80983	549	111	669	220
94	561	134	705	272	0.79835	393	0.78947
95	278	0.80852	424	0.79991	555	114	670
96	0.80991	566	138	706	271	0.78831	388
97	698	274	0.79846	415	0.78981	542	100

(continued)

Mass fraction of ethanol/%	10℃	15℃	20℃	25℃	30℃	35℃	40℃
98	399	0.79975	547	117	684	247	0.77806
99	094	670	243	0.78814	382	0.77946	507
100	0.79784	360	0.78934	506	075	641	203

Appendix 3.2 Density table of water
附录 3.2 水的密度表

Temperature /℃	Density /(kg/m³)	Temperature /℃	Density /(kg/m³)	Temperature /℃	Density /(kg/m³)
1	0.99992	11	0.99963	21	0.99801
2	0.99997	12	0.99953	22	0.99779
3	1.00000	13	0.99941	23	0.99757
4	1.00000	14	0.99928	24	0.99733
5	0.99999	15	0.99913	25	0.99707
6	0.99995	16	0.99898	26	0.99680
7	0.99991	17	0.99881	27	0.99653
8	0.99986	18	0.99862	28	0.99626
9	0.99980	19	0.99843	29	0.99596
10	0.99973	20	0.99823	30	0.99567

Appendix 4 Abbe refractometer
附录 4 阿贝折射仪

An Abbe refractor is an instrument capable of measuring the refractive index N_D and the average dispersion N_F-N_C of a transparent, translucent liquid or solid (mainly to measure transparent liquid). The instrument has a connection to the thermostat. The refractive index N_D at 0-70℃ can be measured by a thermostat. Its measurement range: N_D=1.300-1.700, measurement accuracy of 3×10^{-4}. The instrument is simple to use and fast to obtain data. Abbe refractometer is often used in chemical experiments to determine the composition of binary mixtures.

A4.1 Working principle and structure

The basic principle of abbe refractometer is the law of refraction, see Fig. A4-1.

$$n_1 \sin\alpha_1 = n_2 \sin\alpha_2$$

Where, n_1, n_2 are the refractive index (折射率) of the medium on both sides of the phase interface; and $α_1$ and $α_2$ are incident angles (入射角) and refraction angles (折射角) respectively.

If the light enters the light hydrophobic medium from the light dense medium, the incident angle is less than the refractive angle. Changing the incident angle can make the refractive angle to 90°, and then the incident angle is called the critical angle. The measurement of refractive index is based on the critical angle principle. Observe the light with an eyepiece. It can be seen that the field of view is divided into two parts (see Fig. A4-2). There is a clear boundary between the two, where the bright and dark boundary is the critical angle position (两者之间有明显的分界线，明暗分界处即为临界角位置).

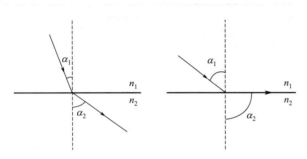

Fig. A4-1　Schematic of refraction law
图附 4-1　折射定律示意图

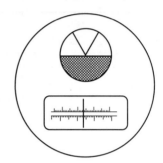

Fig. A4-2　Schematic of the view field of refractor
图附 4-2　折射仪视场示意图

The Abbe refractor can be divided into three types: monocular, binocular and digital type. Although the structure varies, the basic optical principle is the same. The structure of monocular Abbe refractometer is shown in Fig. A4-3.

Fig. A4-3　Abbe refractometer
图附 4-3　阿贝折射仪

1—Reflector (反射镜); 2—Rotation shaft (转轴); 3—Shading panel (遮光板); 4—Thermometer (温度计); 5—Light prism seat (进光棱镜座); 6—Dispersion adjustment handwheel (色散调节手轮); 7—Dispersion value scale circle (色散刻度圈); 8—Eyepiece (目镜); 9—Cover (盖板); 10—Locking wheel (缩紧轮); 11—Refraction prism seat (折射棱镜座); 12—Collecting lens (聚光镜); 13—Thermometer seat (温度计座)

A4.2 Usage（使用方法）

① Constant temperature（恒温）：Place the Abbe refractometer in a sufficient light position, connect the constant temperature water inlet and outlet pipe joint on the light prism seat and refraction prism seat to the super constant temperature tank with rubber pipe, and then adjust temperature control of the heated water bath to the required measurement temperature. After the water bath temperature is stable for 5 minutes, start to use（待水浴温度稳定5分钟后，即可开始使用）.

② Add sample（加样）：The light entry prism holder and refractive prism holder are connected by a rotating shaft. Open the light prism and clean the mirror with a small amount of ether or anhydrous alcohol, wipe the mirror with wipe paper. After the mirror is dried, add the measured liquid to the surface of the refraction prism with a clean dropper, and then cover the light prism. Rotate the lock knob with the hand wheel to fill the field evenly with the liquid layer.

③ Beam focus and adjust（对光和调整）：Open the shading plate, close the reflector, and adjust the eyepiece apparent degree to make the cross image clear. At this point, rotate the handwheel to find the position of the light and dark boundary in the visual field of view. Then rotate the hand wheel to make the dividing line without any color. Fine-tune the hand wheel, so that the boundary is at the center of the cross line. Then rotate the convergent lens appropriately, and the value displayed below the eyepiece field of view is the refractive index of the liquid to be measured（打开遮光板，合上反射镜，调节目镜视度，使十字线成像清晰，此时旋转手轮并在目镜视场中找到明暗分界线的位置，再旋转手轮使分界线不带任何彩色，微调手轮，使分界线位于十字线的中心，再适当转动聚光镜，此时目镜视场下方显示的示值即为被测液体的折射率）.

④ At the end of the measurement, turn off the heated water bath first. Then wipe the surface of the prism clean. If the instrument not used for a longer time, remove the rubber pipe connected to the thermostatic water bath, fully drain the water in the prism thermostatic jacket, and then put the Abbe refractor in the instrument box.

A4.3 Attention points（注意事项）

① The refractive index should be measured at constant temperature, otherwise affecting the test results.

② If the instrument is not used for a long time or the measurement has a deviation, 1-2 drops of naphthalene bromide can be added to the polishing surface of the refraction prism, and then affixed with a standard sample for correction（仪器如果长时间不用或者测量有偏差时，可在折射棱镜的抛光面上加1～2滴溴代萘，再贴上标准试样进行校正）.

③ Keep the instrument clean and do not touch the optical parts with hands. Only use acetone and dimethyl ether to clean the optical components, and then wipe them gently with mirror paper（保持仪器的清洁，严禁用手接触光学零件，光学零件只允许用丙酮、二甲醚等来清洗，并用擦镜纸轻轻擦拭）.

④ The instrument shall avoid strong vibration or impact to prevent damage to the optical parts from affecting their accuracy.

Appendix 4.1 Relationship between refractive index of ethanol-propanol and solution concentration
附录 4.1 乙醇-丙醇折射率与溶液浓度的关系

The relationship between refractive index and mass fraction can be calculated according to the following regression correlation （折射率与质量分率之间的关系可按下列回归关联式计算）：

20℃　$M = 56.60579 - 40.8458 N_D$
25℃　$M = 58.21400 - 42.0170 N_D$
30℃　$M = 58.40500 - 42.1900 N_D$
40℃　$M = 59.28144 - 42.7640 N_D$

Where, M is the mass fraction of ethanol （乙醇的质量分数）; N_D is the refractive index reading.

Note: The above data are measured by experiments.

Appendix 4.2 Relationship between composition and refractive index of n-heptane-methyl cyclohexane system
附录 4.2 正庚烷-甲基环己烷体系的组成与折光率关系

Molar fraction of n-heptane（正庚烷摩尔分数）	Refractive index ［折光率(25℃)］	Molar fraction of n-heptane（正庚烷摩尔分数）	Refractive index ［折光率(25℃)］
0.02	1.4198	0.50	1.4018
0.06	1.4182	0.54	1.4004
0.10	1.4166	0.58	1.3990
0.14	1.4150	0.62	1.3976
0.18	1.4134	0.66	1.3962
0.22	1.4119	0.70	1.3948
0.26	1.4104	0.74	1.3936
0.30	1.4090	0.78	1.3922
0.34	1.4075	0.82	1.3908
0.38	1.4061	0.86	1.3892
0.42	1.4047	0.90	1.3884
0.46	1.4032		

Appendix 5 Gas-liquid equilibrium data
附录 5 气液平衡数据

Appendix 5.1 Gas-liquid equilibrium data of ethanol-aqueous solution at atmospheric pressure (P= 760mmHg)
附录 5.1 常压下乙醇-水溶液气液平衡数据

Molar fraction of ethanol in liquid phase (液相中乙醇的摩尔分数)	Molar fraction of ethanol in gas phase (气相中乙醇的摩尔分数)	Molar fraction of ethanol in liquid phase (液相中乙醇的摩尔分数)	Molar fraction of ethanol in gas phase (气相中乙醇的摩尔分数)
0.0	0.0	0.45	0.635
0.01	0.11	0.50	0.657
0.02	0.175	0.55	0.678
0.04	0.273	0.60	0.698
0.06	0.340	0.65	0.725
0.08	0.392	0.70	0.755
0.10	0.430	0.75	0.785
0.14	0.482	0.80	0.82
0.18	0.513	0.85	0.855
0.20	0.525	0.894	0.894
0.25	0.551	0.90	0.898
0.30	0.575	0.95	0.942
0.35	0.595	1.0	1.0
0.40	0.614		

Appendix 5.2 Gas-liquid equilibrium data of ethanol-propanol at atmospheric pressure (molar fraction)
附录 5.2 常压下乙醇-丙醇气液平衡数据(摩尔分率)

No.	1	2	3	4	5	6	7	8	9	10	11
$t/℃$	97.60	93.85	92.66	91.60	88.32	86.25	84.98	84.13	83.06	80.59	78.38
x	0	0.126	0.188	0.210	0.358	0.416	0.546	0.600	0.663	0.844	1.0
y	0	0.240	0.318	0.349	0.550	0.650	0.711	0.760	0.814	0.914	1.0

Note: The above equilibrium data are taken from: J. Gembling, U. Onken, Vapor-Liquid Equilibrium Data Collection-Organic Hydroxy Compounds: Alcohol (p.336).

Appendix 5.3 Gas-liquid equilibrium data of n-heptane methyl cyclohexane at atmospheric pressure
附录 5.3 常压下正庚烷-甲基环己烷的气液平衡数据

Molar fraction of n-heptane in liquid phase (液相中正庚烷的摩尔分数)	Molar fraction of n-heptane in gas phase (气相中正庚烷的摩尔分数)	Molar fraction of n-heptane in liquid phase (液相中正庚烷的摩尔分数)	Molar fraction of n-heptane in gas phase (气相中正庚烷的摩尔分数)
0.031	0.035	0.559	0.578
0.058	0.062	0.599	0.618
0.095	0.103	0.6407	0.666
0.133	0.143	0.709	0.728
0.18	0.192	0.756	0.771
0.216	0.229	0.796	0.81
0.2715	0.289	0.843	0.8535
0.307	0.333	0.879	0.89
0.363	0.381	0.906	0.913
0.401	0.42	0.913	0.94
0.456	0.475	0.954	0.9625
0.501	0.521	0.98	0.986

Appendix 6 Characteristic parameters of the four packings
附录 6 四种填料的特性参数

Packing name (填料名称)	Porcelain Raschig ring (瓷拉西环)	Metal θ ring (金属 θ 环)	Metal wave wire mesh (金属波纹丝网)	Star packing (plastic) [星型填料(塑料)]
Specifications (diameter×height×thickness)/mm	12×12×1.3	10×10×0.1	CY-type	15×8.5×0.3
Specific surface area $a_t/(m^2/m^3)$	403	540	700	850
Void ratio $\varepsilon/(m^3/m^3)$	0.764	0.97	0.85	
Packing factor $a_t/\varepsilon/(m^2/m^3)$	903			

Appendix 7 Oxygen dissolving instrument
附录 7 溶氧仪

A7.1 Basic structure

The structure of the oxygen probe is illustrated in Fig. A7-1.

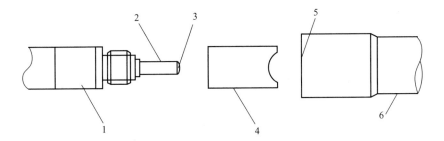

Fig. A7-1　Schematic of oxygen probe structure
图附 7-1　氧探头的结构示意

1—Temperature compensator（温度补偿器）；2—Silver electrode（银电极）；3—Gold electrode（金电极）；4—Membrane fixator（膜固定器）；5—Spongy body（kept wet）［海绵体（保持湿润）］；6—Protective case（保护套）

A7.2　Directions（使用方法）

(1) For the first use

① Install the battery, plug the oxygen probe and the temperature probe on the instrument.

② Prepare a glass of water and sit in the air for hours to make it a saturated oxygen solution.

③ Insert the oxygen probe and the temperature probe into the saturated oxygen water solution simultaneously, polarizing it for about 10min.

④ Keep the probe connected all the time, no longer requires a polarization operation.

(2) Measurement

① Press ON to turn on the instrument.

② Insert the two probes into the saturated oxygen solution at the same time, and keep the solution flowing.

③ Presses MODE to display "%" in the lower right corner of the screen, and adjust the Slope knob to make the data up to 100% in the screen.

④ Press the MODE key to display zero in the lower left corner and adjust the zero key to zero data on the screen.

⑤ Repeat ③ and ④ to keep the full degree at 100% when the zero indicator is zero.

⑥ Put the probe into the tested solution while requiring stirring.

⑦ Press MODE to display mg/L in the upper right corner, and the data in the screen is the oxygen content of the solution (mg/L).

(3) Attention points（注意事项）

① After measuring, press the "OFF" button to turn off the instrument and do not remove the battery and probe.

② Insert the oxygen probe into the protective case containing sufficient water.

③ When using the probe and instrument, handle gently and pay special attention to not making the membrane of the oxygen probe collide with other hard objects so as not to break the membrane.

④ The measuring range of the instrument is aqueous solution containing 0-19.9mg/L oxygen, the measuring temperature is －30-150℃.

A7.3 Oxygen concentrations in water at different temperatures（不同温度的氧在水中的浓度）

Temperature/℃	Concentration/(mg/L)	Temperature/℃	Concentration/(mg/L)
0.00	14.6400	18.00	9.6827
1.00	14.2453	19.00	9.4917
2.00	13.8687	20.00	9.3160
3.00	13.5094	21.00	9.1357
4.00	13.1668	22.00	8.9707
5.00	12.8399	23.00	8.8116
6.00	12.5280	24.00	8.6583
7.00	12.2305	25.00	8.5109
8.00	11.9465	26.00	8.3693
9.00	11.6752	27.00	8.2335
10.00	11.4160	28.00	8.1034
11.00	11.1680	29.00	7.9790
12.00	10.9305	30.00	7.8602
13.00	10.7027	31.00	7.7470
14.00	10.4838	32.00	7.6394
15.00	10.2713	33.00	7.5373
16.00	10.0699	34.00	7.4406
17.00	9.8733	35.00	7.3495

Appendix 8　752-model spectrophotometer
附录8　752-型分光光度计

The 752-model spectrophotometer is a UV grating spectrophotometer with a determination wavelength of 200-800nm. Using holographic sparkles grating monochromator, it has the advantages of high wavelength accuracy, good monochromator and low stray light. It can automatically switch tungsten halogen lamps and deuterium lamps. It features a 3-digit LED digital display with T.A.C（transmittance absorbancy concentration）direct reading function. Equipped with analog recorder interface, it can be used for kinetic energy reaction and other special tests. Sample chamber can accommodate 0.5-5cm cuvette.

A8.1　Structure principle（结构原理）

752-model spectrophotometer is composed of light source room（光源室），monochromator（单色器），sample chamber（样品室），photocell cartridge（光电管暗盒），electronic system and digital display, etc. The working principle of the instrument is as follows: The continuous radiation from a tungsten lamp or hydrogen lamp is directed to the monochromator inlet slit after being selected by a filter and concentrated by collecting lens. This slit is located exactly in the focal plane of the spotlight mirror and the monochromatic internal collimator mirror, so the composite light entering the monochromator is reflected by the plane mirror and the collimation mirror, and becomes parallel light to the dispersion grating. By diffraction, the incident composite light through the grating forms a continuous monochromatic spectrum uniformly arranged in a certain order. The monochromatic spectrum then returns to the collimating mirror and is imaged on the exit slit by focusing principle. The monochromatic light of a specified wide band is selected from the exit slit and falls into the center of the sample chamber through the spotlight mirror. After being absorbed by the sample, the transmitted light is transmitted to the photocell cathode through the photogate. According to the principle of the photoelectric effect, a weak photocurrent is generated. The amplified photoelectric current by a current amplifier is sent to a digital display to measure transmittance and absorbance, or show the concentration C of the tested sample through the logarithmic amplifier.

A8.2　Directions（使用方法）

① Set the sensitivity knob to "1" (minimum magnification).

② Turn on the power switch, light on the tungsten lamp, and preheat for 30 min. If ultraviolet light is required, turn on the "hydrogen light" switch, then press the hydrogen light trigger button to light the hydrogen light and warm up for 30min.

③ Set the selector switch to "T".

④ Open the sample chamber cover and adjust the 0% knob to display the number as "0.000".

⑤ Adjust the wavelength knob and select the required wavelength.

⑥ Put cuvettes with reference solution and measured solution into the cuvettes tray.

⑦ Cover the sample chamber, make the light path passes through the reference solution cuvette, and adjust the light transmittance knob, so that the number is displayed as 100.0% (T). If the display is less than 100.0% (T), appropriately increase the sensitivity. Then put the measured solution in the optical path, and the digital display value is the light transmittance of the measured solution.

⑧ If the transmittance is not required, adjust the selection switch to "A" after the instrument also displays 100.0% (T). Then put the measured solution in the optical path, and the digital display value is the absorbance of the solution.

⑨ If the selection switch is set to "C", place the solution with known calibration concentration in the optical path, adjust the temperature knob to make the digital display as the calibration value, and then place the measured solution in the optical path, the corresponding concentration value will be displayed.

A8.3　Attention points（注意事项）

① The glass cuvette can be used for determine wavelength above 360nm; A quartz cuvette is used for wavelength below 360nm（测定波长在 360nm 以上时，可用玻璃比色皿；波长在 360nm 以下时，要用石英比色皿）. Suck dry the outside of the cuvette with absorbent paper, not to touch the smooth surface by hand.

② Do not replace the matched cuvette with other instruments. Supplements should only be used after calibration.

③ When opening and closing the sample chamber cover, be careful to prevent damage to the optical gate.

④ When not measuring, the sample chamber cover should be in open state. Otherwise the photocell fatigue will lead to digital display instability.

⑤ When the light wavelength adjustment range is large, it takes a few minutes to work. The reason is that the photocell takes some time to respond to light.

⑥ Keep the instrument dry and clean.

Appendix 9　Principles of automatic acquisition of computer data and automatic control
附录 9　计算机数据自动采集及自动控制原理

The so-called computer collection is to convert some physical quantities in the project, such as temperature, pressure and flow, into DC signals through sensors. These signals are amplified and converted into 0-5V DC voltage signals, and then converted into digital inputs to computers through A/D converter. Finally, the collected physical quantity are displayed or calculated, drawn and so on through computer programming. The function of the signal processing amplifier: ① Transform the weak electrical signals into a DC voltage of 0-5V; ②Inhibit interference and reduce noise. A DC signal of 0-5V input from the signal processing amplifier is usually called an analog quantity（模拟量）, which can be expressed by infinite long numbers, such as 3.5281…（V）. But by processing these analog quantities with a computer or other digital processor, we can only deal with quantities of finite length, which is called digital quantities（数字量）. Therefore, the analog quantity must convert into a digital quantity. This conversion device is called the analog/digital converter, known as the A/D converter（A/D 转换器）.

The automatic control system consists of measured coefficient, adjustment object, automatic regulator, sensor and other components.

According to the process conditions of the measured object, the given value （within the computer software） is transmitted to the intelligent instrument by the computer through the communication interface. The intelligent instrument compares the given value with the measured parameters. When there is a deviation, the opening of the adjustment valve is automatically changed and the flow is adjusted so that the deviation between the measured parameters and the given value is within the allowed range.

The circuit diagram of pump computer data acquisition automatic control system is as follows (Fig. A9-1):

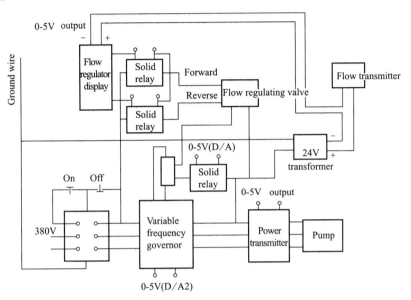

Fig. A9-1 Circuit diagram of pump computer data acquisition automatic control system
图附 9-1 泵计算机数据采集自控系统电路示意图

Appendix 10 Installation and operation instructions of "Chemical Engineering Principle Experiment CAI"
附录 10 《化工原理实验 CAI》安装操作及使用说明

A10.1 Installation instructions of "Chemical Engineering Principle Experiment CAI"(《化工原理实验 CAI》安装操作说明)

Welcome to the multimedia simulation CAI courseware of the chemical principle experiment developed by the College of Environmental Science and Engineering, Donghua University. There are still some shortcomings. We sincerely hope you will offer your valuable advice.

Recommended best configuration: Windows 7/10 operating system; 500 MB hard disk space, 1024×768 resolution (standard small font).

Taking the comprehensive experiment of fluid mechanics as an example, the installation steps are as follows (现以流体力学综合实验为例，安装步骤如下):

(1) The following screen appears after running "setup.exe" in the CD, please click "Next" button to enter the next installation process, and click "Cancel" to exit the installation process (Fig. A10-1).

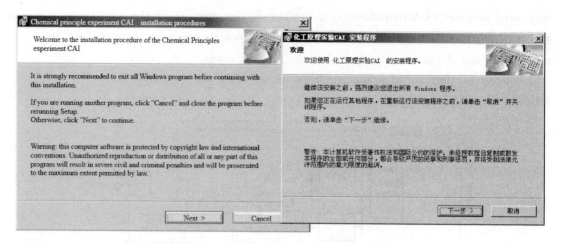

Fig. A10-1 Set up interface of "Chemical Principles Experiment CAI"

图附 10-1 《化工原理实验 CAI》——设置界面

(2) After the following screen appears, please read the license agreement (Fig. A10-2). You can select "I agree to the terms of this License Agreement" and then click the "Next" button to proceed to the next installation step. If you are in doubt about the license agreement, you can select "I do not agree with the terms of the license agreement" and click "Cancel" to exit the installation process.

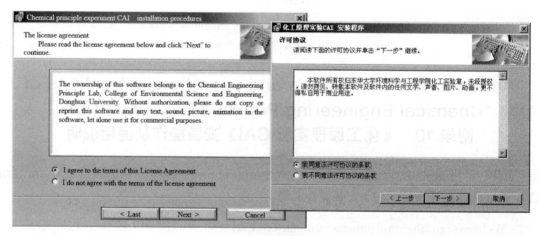

Fig. A10-2 License agreement of "Chemical Principles Experiment CAI"

图附 10-2 《化工原理实验 CAI》——许可协议

(3) After agreeing to the terms of the license agreement, click "Next" and the following screen will appear (Fig. A10-3). Enter relevant user information and click "Next" to enter the next installation process.

(4) After browsing the basic information of the software shown in the figure below (Fig. A10-4), click the "Next" button to enter the next installation process.

(5) After the following screen appears (Fig. A10-5), you can change the installation directory or use the default installation path, and then click the "Next" button to enter the next installation process.

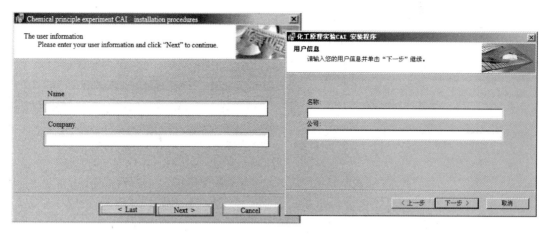

Fig. A10-3　User information of "Chemical Principles Experiment CAI"

图附 10-3　《化工原理实验 CAI》——用户信息

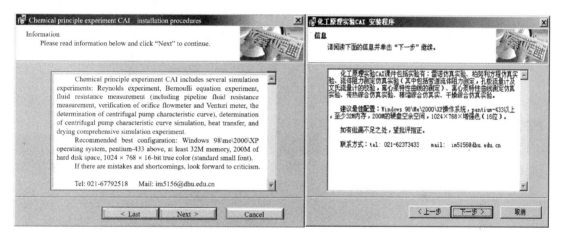

Fig. A10-4　Basic information on the software of "Chemical Principles Experiment CAI"

图附 10-4　《化工原理实验 CAI》——软件基本信息

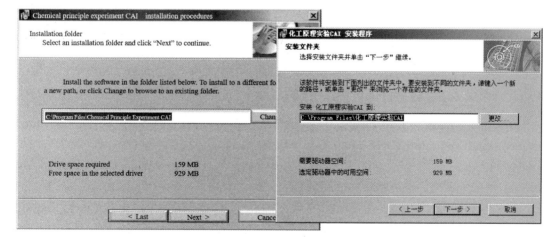

Fig. A10-5　Installation path of "Chemical Principles Experiment CAI"

图附 10-5　《化工原理实验 CAI》——安装路径

(6) After the following screen appears (Fig. A10-6), click "Next" button to install the software of chemical engineering principle experiment CAI.

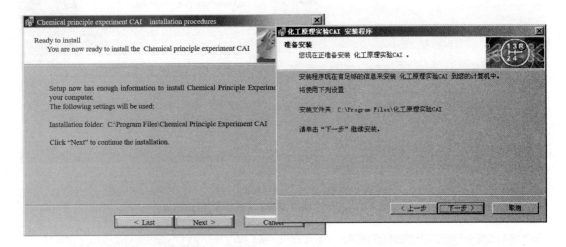

Fig. A10-6   Installation of "Chemical Principles Experiment CAI"
图附 10-6 《化工原理实验 CAI》——安装

(7) The progress bar shows that installation is in progress (Fig. A10-7).

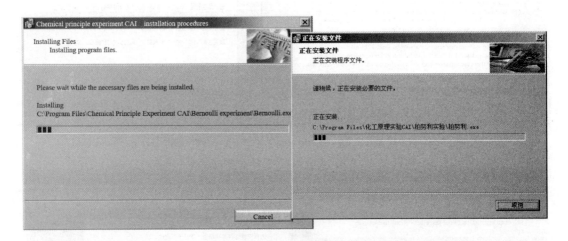

Fig. A10-7   Installation progress of "Chemical Principles Experiment CAI"
图附 10-7 《化工原理实验 CAI》——安装进度

(8) Finally, the following interface appears (Fig. A10-8). Click "Finish" button to complete the installation of the program.

(9) Thank you for installing the "Chemical Principles Experiment CAI". Next, you can find "Chemical Engineering Principle Experiment CAI" in the "Start-Program-Chemical Engineering Principle Experiment CAI" menu (as shown on the left of Fig. A10-9). Simply click the "Chemical Engineering Principle Experiment CAI" icon on the desktop (as shown on the right of the figure below) to run the Chemical Engineering Principle Experiment CAI.

Appendix

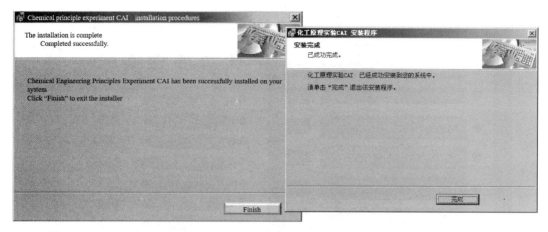

Fig. A10-8 Installation complete of "Chemical Principles Experiment CAI"
图附 10-8 《化工原理实验 CAI》——完成安装

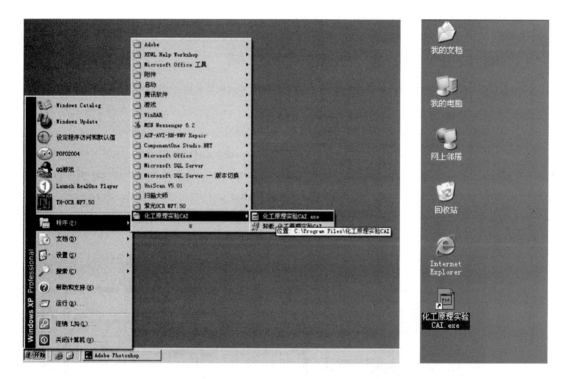

Fig. A10-9 Find menu of "Chemical Principles Experiment CAI"
图附 10-9 《化工原理实验 CAI》——查找菜单途径

A10.2 Operational approach of "unit operation experiment CAI"(《化工原理实验 CAI》操作方法)

(1) Startup simulation experiment（启动仿真实验）

Click "Start-Program-Chemical Engineering Principles Experiment CAI—Chemical Engineering Principles Experiment CAI" with the mouse to start the experiment. After opening animation，the following selection interface will appear（Fig. A10-10）：

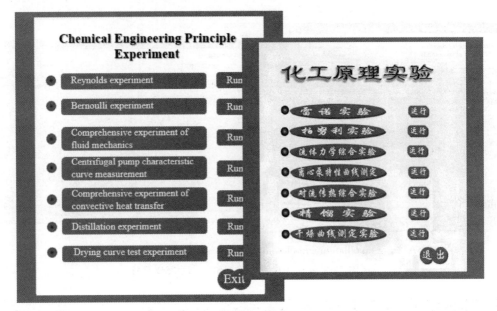

Fig. A10-10　Initiate simulation experiments
图附 10-10　启动仿真实验

Move the mouse over the experiment name, and click the left mouse button to enter the experiment.

（2）System functions（系统功能的使用）

Considering the following example focusing on measurement of fluid resistance in smooth tube, the common system functions of this software, as well as the related adjustment or use methods, are introduced below. Other experimental procedures are similar and will not be described in this article.

The main interface is as follows（Fig. A10-11）:

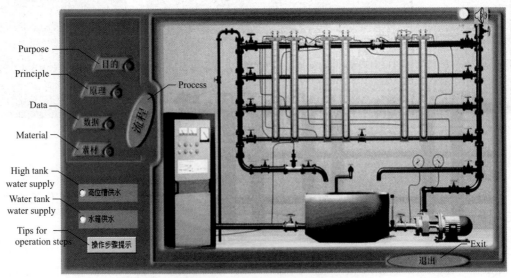

Fig. A10-11　Main interface (example of fluid resistance measurement in a smooth tube)
图附 10-11　主界面（以光滑管中流体阻力测定为例）

Appendix

The interface is divided into three parts: Menu and corresponding selection area (left), experiment demonstration area (right picture area), and experiment exit key (bottom right).

The left menu is the system call menu, with five items of "experimental purpose, experimental principle, experimental process, experimental data processing, and related materials", which are respectively described as (Fig. A10-12):

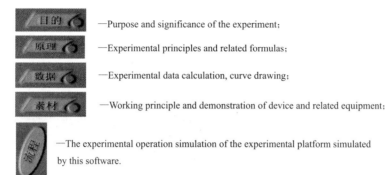

—Purpose and significance of the experiment;

—Experimental principles and related formulas;

—Experimental data calculation, curve drawing;

—Working principle and demonstration of device and related equipment;

—The experimental operation simulation of the experimental platform simulated by this software.

Fig. A10-12　System menu in the main interface

图附 10-12　主界面中系统菜单

To start one of the above windows, move your mouse over the corresponding item and left click. Click the Exit button at the lower right to return to the previous menu option.

A10.3　Method of "Data processing"（"数据处理"使用方法）

The main interface is as follows:

Original data　—The experimental test data can be input in the corresponding data input box. The values entered in this interface are the software defaults, which can be changed (Fig. A10-13).

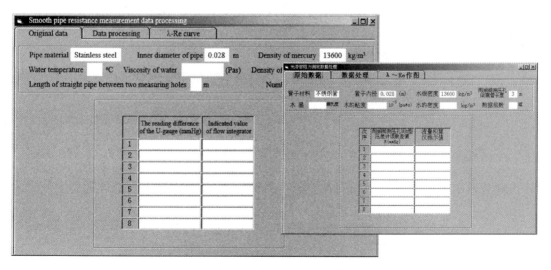

Fig. A10-13　Raw data interface

图附 10-13　原始数据界面

Data processing —Click the "Data processing" button to obtain the experimental processing results (Fig. A10-14).

Note: Be sure to enter a reasonable value in the "Original data" area, enter the correct number of corresponding groups in **Number of data sets** to get the correct result. Otherwise, the system may jump out. If the system jumps out, use the "ctrl+alt+delete" combination key to close the simulation program that is still running in the background, and then re-enter the experiment.

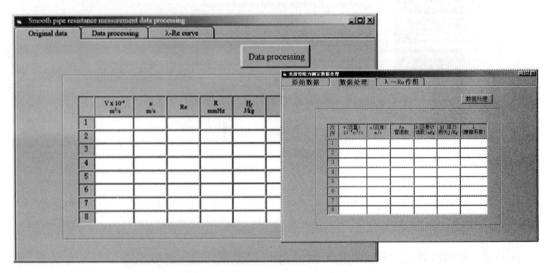

Fig. A10-14 Data processing interface
图附 10-14 数据处理界面

λ-Re curve —After obtaining the experimental data calculation results in the data processing area, click "Plotting" here to get the curve graph, and click "Return" to exit the data processing interface and return to the main interface (Fig. A10-15).

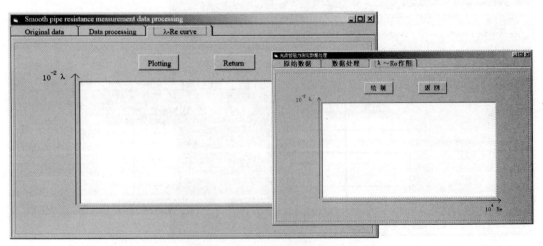

Fig. A10-15 Plotting curve interface
图附 10-15 绘制曲线界面

A10.4 Experimental operation process simulation and the use of related simulation equipment

① Experimental platform（实验平台）(Fig. A10-16)

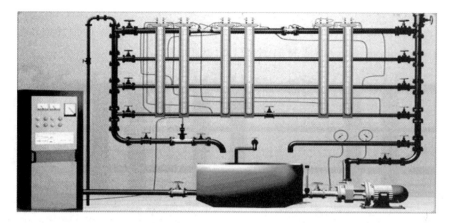

Fig. A10-16　Simulation interface of experimental operation flow—Experimental platform
图附 10-16　实验操作流程仿真界面——实验平台

② Local enlarged part（局部放大的部件）　When the mouse moves to the part with the zoom picture, the mouse pointer becomes a small hand. Press the left mouse button at the time to display the local enlarged image, and click the left mouse button on the local zoom picture to turn off it.

The local enlarged diagram includes: orifice flowmeter, Venturi flowmeter, gate valve, U-type differential pressure meter, pressure gauge, vacuum meter, liquid crystal meter of flow integrator, and control button of flow integrator, etc.

③ Flow control（流量控制）　Sliding the slider shown in Fig. A10-17 can change the magnitude of the flow. The experimental procedure is for simulation, and the data obtained

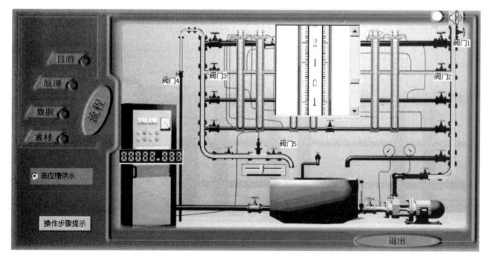

Fig. A10-17　Simulation interface of experimental operation—Flow rate control
图附 10-17　实验操作流程仿真界面——流量控制

are not completely consistent with the reality, but the law of flow change is reasonable.

④ Error-correction processing (纠错处理) In case of wrong choice, the system will give a prompt. You can complete the experiment smoothly according to the prompt, at the same time, this process is equipped with voice explanation, and text explanation. People who first use the software, please listen carefully or read the operation steps tips, and then simulate the experiment, which can yield twofold results with half the effort.

⑤ Material presentation (素材演示) (Fig. A10-18)

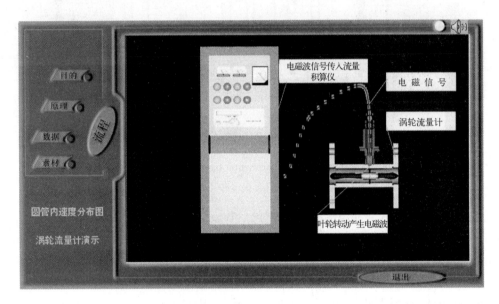

Fig. A10-18 Simulation interface of experimental operation—Material demonstration
图附 10-18 实验操作流程仿真界面——素材演示

Select the material to see in the lower left, such as turbine flowmeter demonstration. Left-click, and the relevant 3D animation demonstration will appear in the right demonstration area.

Pipeline resistance measurement includes: velocity distribution diagram in round pipe, turbine flowmeter demonstration, gate valve, valve core, valve shell.

The calibration of orifice flowmeter and Venturi flowmeter includes: orifice flowmeter, Venturi flowmeter.

Determination of characteristic curve of centrifugal pump includes: pressure gauge, pump principle, pump shell, impeller, bottom valve, air binding of pump and pump cavitation phenomenon.

⑥ Sound on and off (声音开启与关闭) On the upper right corner of the interface containing sound, there will be 🔊 or 🔇 logo. Left-click the 🔊 logo to close the sound being played at the current interface; left-click the 🔇 logo to open the sound contained in the current interface.

Note: the other experiment CAI can refer to this experiment operation.

Appendix 11　Data recording and collation table of unit operation experiment
附录 11　化工原理实验数据记录及整理表格

Appendix 11.1　Data recording and collation table of Reynolds experiment
附录 11.1　雷诺实验数据记录及整理表格

Orifice aperture（孔板孔径）_____；Flow coefficient（流量系数）_____；
Water temperature（水温）_____；Density of water（水的密度）_____；
Viscosity of water（水的黏度）_____.

No.（序号）	U-tube reading of orifice meter, R（孔板流量计读数）/mmH$_2$O	Flow rate V（流量）/(m^3/h)	Reynolds number Re（雷诺数）	Phenomenon（现象）	Flow pattern（流型）

Appendix 11.2　Data recording and collation table of Bernoulli equation experiment
附录 11.2　伯努利方程实验数据记录及整理表格

Opening of flow control valve（流量调节阀的开度）	Cross-section 1（截面 1）		Cross-section 2（截面 2）		Cross-section 3（截面 3）		Cross-section 4（截面 4）	
	Hydrostatic head（静压头）	Total head（冲压头）	Hydrostatic head（静压头）	Total head（冲压头）	Hydrostatic head（静压头）	Total head（冲压头）	Hydrostatic head（静压头）	Total head（冲压头）
Fully open, H/mmH$_2$O（全开）								
Half open, H'/mmH$_2$O（半开）								

Appendix 11.3 Data recording and collation table of the pipeline fluid resistance determination
附录11.3 管道流体阻力的测定实验数据记录及整理表格

(1) Experimental data recording（实验数据记录）

① Measurement of straight pipe resistance（直管阻力测定）

Pipe material（管子材料）_____ Diameter of pipe（管子内径）_____

Density of mercury（水银密度）_____ Water temperature（水温）_____

Viscosity of water（水的黏度）_____ Density of water（水的密度）_____

Length of straight pipe between two measuring ports（两闸阀测压孔间直管长度）_____

No. (序号)	Reading of U-tube manometer between two measuring ports （两闸阀测压孔间U形压差计读数）			Indicated value of flow integrator （流量积算仪指示值）
	Left /mm	Right /mm	Difference value /mm	

② Determination of equivalent length l_e of No.1 gate valve (full opening) [1# 闸阀（全开时）的当量长度 l_e 测定]

Pipe material（管子材料）_____ Diameter of pipe（管子内径）_____

Density of mercury（水银密度）_____ Water temperature（水温）_____

Viscosity of water（水的黏度）_____ Density of water（水的密度）_____

Length of straight pipe between two measuring ports L_1（闸阀测压孔间直管长度 L_1）_____

Length of straight pipe between two measuring ports L_2（闸阀测压孔间直管长度 L_2）_____

No. (序号)	Total reading of U-tube manometer between measuring ports （闸阀总测压孔间U形压差计读数）			Reading of U-tube manometer between measuring ports （闸阀测压孔间U形压差计读数）			Indicated value of flow integrator （流量积算仪指示值）
	Left /mm	Right /mm	Difference value /mm	Left /mm	Right /mm	Difference value /mm	

(2) Experimental data collation (实验数据整理)

① Measurement of straight pipe resistance (直管阻力测定)

No. (序号)	Flow rate (流量), $V/(m^3/s)$	Velocity (流速), $u/(m/s)$	Reynolds number (雷诺数), Re	U-tube manometer reading (压差计读数), $R/(mmHg)$	Resistance loss (阻力损失), $h_f/(J/kg)$	Fraction coefficient (摩擦系数), λ

② Determination of equivalent length l_e of No. 1 gate valve (full opening) [1# 闸阀 (全开时) 的当量长度 l_e 测定]

No. (序号)	V /(m³/s)	u /(m/s)	Total resistance (总阻力), $h_{f1}/(J/kg)$	Resistance (阻力), $h_{f2}/(J/kg)$	Gate resistance (闸阀阻力), $H'_f/(J/kg)$	Straight pipe (直管阻力), $h_f/(J/kg)$	λ	l_e/m

Mean l_e = _____

Appendix 11.4 Data recording and collation table of calibration experiment of orifice and Venturi flowmeter
附录 11.4 孔板流量计及文丘里流量计的校验实验数据记录及整理表格

(1) Experimental data recording (实验数据记录)

Pipe material (管子材料) _____ Diameter of pipe (管子内径) _____
Density of mercury (水银密度) _____ Water temperature (水温) _____
Density of water (水的密度) _____ Orifice aperture (孔板孔径) _____
Diameter of Venturi pipe (文丘里管管径) _____

No. (序号)	Reading of U-tube manometer for orifice flowmeter (孔板流量计处 U 形压差计读数) /mmHg			Reading of U-tube manometer for Venturi flowmeter (文丘里流量计处 U 形压差计读数) /mmHg			Indicated value of flow integrator (流量积算仪指示值)
	Left /mm	Right /mm	Difference value /mm	Left /mm	Right /mm	Difference value /mm	

(2) Experimental data collation（实验数据整理）

No.（序号）	Flow rate $V \times 10^{-4}$ /(m³/s)	Orifice flowmeter（孔板流量计)			Venturi flowmeter（文丘里流量计)		
		Velocity u_1/(m/s)	Reading of U-tube meter （压差计读数），R/mmHg	Hydrostatic head loss （静压能损失），Δh_1/(J/kg)	Velocity u_2/(m/s)	Reading of U-tube meter （压差计读数），R/mmHg	Hydrostatic head loss （静压能损失），Δh_2/(J/kg)

Appendix 11.5　Data recording and collation table of determination of centrifugal pump characteristic curve
附录 11.5　离心泵特性曲线测定实验数据记录及整理表格

(1) Determination experiment of characteristic curve of centrifugal pump（离心泵特性曲线测定实验）

① Experimental data recording（实验数据记录）

Water temperature（水温）_____　　Density of water（水的密度）_____

η of motor（电机效率）_____　　Vertical distance between the pressure gauge and the vacuum gauge（真空表与压力表的高度差）h_0_____

No.（序号）	Vacuum pressure （真空表读数） /mmHg	Gauge pressure （压力表读数） /(kgf/cm²)	Power of motor （功率表读数） /kW	Indicated value of flow integrator （流量积算仪指示值）

② Experimental data collation（实验数据整理）

No.（序号）	Flow rate, $V \times 10^{-4}$ /(m³/s)	Vacuum gauge （真空表） H/mH₂O	Pressure gauge （压力表）, H/mH₂O	H /mH₂O	N_{shaft} /kW	N_e /kW	η /%

Note：1mmH₂O=9.80665Pa．

(2) The experimental data of pipeline characteristic curve measurement (管路特性曲线测定实验数据)

No.（序号）	Motor frequency（电机频率），/Hz	Motor speed（电机转速）/(r/min)	Δp of the flowmeter（流量计压差）/kPa	Pump outlet pressure（泵出口压力）/MPa	Pump inlet pressure（泵入口压力）/MPa	Pump head（压头）/m	Flow rate /(m³/h)

Appendix 11.6 Data recording and collation table of comprehensive heat transfer experiment
附录 11.6 传热综合实验数据记录及整理表格

(1) Experimental data recording (实验数据记录)

Device No. _____   Inner diameter of heat transfer pipe _____

Pipe length of heat transfer _____

No.（序号）	Ordinary double-pipe（普通套管）				Enhanced double-pipe（强化套管）			
	ΔP of the flowmeter/kPa	T(inlet)/℃	T(outlet)/℃	E/mV	ΔP of the flowmeter/kPa	T(inlet)/℃	T(outlet)/℃	E/mV

(2) Experimental data collation (实验数据整理)

① Ordinary double-pipe heat exchanger (enhanced double-pipe heat exchanger)

No.（序号）	Qualitative temperature（定性温度），$\bar{t}$/℃	Air flow rate V_t/(m³/h)	Air flow rate V/(m³/h)	Air density ρ/(kg/m³)	Air flow rate（空气流量）W/(kg/h)	Specific heat of air C_p/[kJ/(kg·℃)]	$\Delta t = t_2 - t_1$/℃	Heat transfer（传热量）Q/W

(continued)

Mean temperature of tube wall, t_w/℃	LMTD Δt_m /℃	Heat transfer area A_i/m²	α_i (传热膜系数)/[W/(m²·℃)]	Air velocity u/(m/s)	μ /Pa·s	λ/[W/(m·℃)]	Re (雷诺数)	Pr	Nu	$Nu/Pr^{0.4}$

② Enhanced heat transfer effect

No. (序号)	Reynolds number (雷诺数)Re	Nusselt number (努塞特数)Nu_0	Nu	Nu/Nu_0

Appendix 11.7 Data recording and collation table of comprehensive experiment of heat transfer coefficient K measurement
附录 11.7 换热器传热系数 K 值测定综合实验数据记录及整理表格

(1) Experimental data recording（实验数据记录）

Name of heat exchanger（换热器名称）_____

Heat transfer area（传热面积）_____ m²

Flow mode (流动方式)	No. (序号)	Cold fluid(冷流体)			Hot fluid(热流体)		
		Meter reading (流量计读数)	t_1(inlet) /℃	t_2(outlet) /℃	Meter reading (流量计读数)	T_1(inlet) /℃	T_2(outlet) /℃
Parallel flow (并流)							
Countercurrent (逆流)							

(2) Experimental data collation（实验数据整理）

Flow mode	No.（序号）	Qualitative temperature（定性温度）τ/℃	Volume flow, V /(m³/h)	Density ρ /(kg/m³)	Mass flow, W /(kg/h)	Specific heat, C_p /[kJ/(kg·℃)]	Δt /℃	Heat transfer rate（传热量）Q/W	LMTD, Δt_m /℃	K /[W/(m²·℃)]
Parallel flow										
Counter-current										

Appendix 11.8 Data recording and collation table of plate column distillation experiment
附录 11.8 板式塔精馏实验数据记录及整理表格

Distillated tower type（精馏塔型）_____
Actual tower plate number（实际塔板数）_____
Experimental condition（实验条件）_____

No.（序号）	Tower top sample（塔顶样品）				Tower bottom sample（塔底样品）				Plate efficiency（塔板效率）		理论板数 N_T
	t/℃	ρ or N_D（密度或折射率）	Mass fraction/%	Mole frac.（摩尔分率）	t/℃	ρ or N_D	Mass fraction/%	Mole frac.	Over-all（全塔）/%	Murphree（单板）/%	

Appendix 11.9 Data recording and collation table of packed column distillation experiment
附录 11.9 填料塔精馏实验数据记录及整理表格

No.（序号）	Tower top sample（塔顶样品）			Tower bottom sample（塔底样品）			Reflux ratio（回流比）	N_T（理论板数）	HETP（等板高度）/m
	t/℃	Refractive index（折射率）	Composition	t/℃	Refractive index（折射率）	Composition			

Appendix 11.10 Data recording and collation table of determination of hydromechanical properties of plate column
附录 11.10 板式塔流体力学性能测定数据记录及整理表格

(1) Experimental data recording（实验数据记录）

No. (序号)	Water flow （水流量） /(L/h)	Gas flow （气流量） /(m³/h)	Pressure drop （压降）		Leakage （漏液量） /kg	Time （时间） /s	Entrainment amount （雾沫夹带量） /kg
			Left/Pa	Right/Pa			

(2) Experimental data collation（实验数据整理）

No. (序号)	Gas flow （气速）/(m/s)	Pressure drop （压降）/Pa	Leakage rate （漏液率）/%	Entrainment rate （雾沫夹带率）/%

Appendix 11.11 Data recording and collation table of absorption-desorption experiment
附录 11.11 吸收-解吸实验数据记录及整理表格

(1) Basic data（基本数据）

Desorption tower diameter（解吸塔径）$\varphi = 0.1$m；

Absorption tower diameter（吸收塔径）$\varphi = 0.032$m；

The height of packing layer（填料层高度）0.8m；

Metal θ ring, metal corrugated wire mesh packing（金属 θ 环、金属波纹丝网填料）；

p_0（atmospheric pressure）$= 101.3$kPa；Gas temperature _____ ℃；

The water temperature is _____ ℃；Spray amount（喷淋量）L：_____ L/h.

(2) Measurement of hydrodynamic performance of packing tower（填料塔流体力学性能测定）

① Pressure drop determination of dry packing（干填料压降测定）

No. (序号)	V_1 /(m³/h)	P_{gas}/Pa			P_{tower}/Pa			V_2 /(m³/h)	u /(m/s)
		Left	Right	ΔP_{gas}	Left	Right	$\Delta P_{tower}/Z$ (Pa/m)		

② Pressure drop determination of wet packing（湿填料压降测定）

No. (序号)	V_1 /(m³/h)	P_{gas}/Pa			P_{tower}/Pa			V_2 /(m³/h)	u /(m/s)
		Left	Right	ΔP_{gas}	Left	Right	$\Delta P_{tower}/Z$ (Pa/m)		

（3）Determination of overall liquid mass transfer coefficient $K_x a$ and total liquid mass transfer unit height H_{oL}（液相体积总传质系数 $K_x a$ 及液相总传质单元高度 H_{oL} 的测定）

① Experimental data recording（实验数据记录）

No. (序号)	Spray volume (喷淋量), L/(L/h)	Solution concentration (溶液浓度) /(kmol/m³)	Left	Right	Pressure drop of packed tower (填料塔压降), ΔP_{tower}/kPa	Liquid oxygen concentration at the top of the tower (塔顶液相氧浓度) /(mg/L)	Liquid oxygen concentration at the bottom of the tower (塔底液相氧浓度) /(mg/L)

② Experimental data collation（实验数据整理）

No.	E /kPa	P /kPa	x_1	X_2	x_1^*	X_2^*	Δx_m	G_A /[kmol/(m²·h)]	V_p /m³	$K_x a$ /[kmol/(m²·h)]	H_{oL} /m

E：Henry coefficient（亨利系数）；
P：Total system pressure（系统总压）；
x_1：Molar fraction of inlet liquid phase（液相进塔浓度摩尔分数）；
X_2：Molar fraction of outlet liquid phase（液相出塔浓度摩尔分数）；
x_1^*：Equilibrium liquid concentration with inlet gas phase y_1（与进塔气相 y_1 平衡的液相浓度）；
X_2^*：Equilibrium liquid concentration with outlet gas phase y_2（与出塔气相 y_2 平衡的液相浓度）；
Δx_m：Logarithmic mean concentration difference of liquid phase（液相对数平均浓度差）；
G_A：Oxygen desorption volume per unit time（单位时间氧解吸量）；
V_p：Packing volume（填料层体积）；
$K_x a$：Overall mass transfer coefficient of the liquid volume（液相体积总传质系数）；
H_{oL}：Height of total mass transfer unit in liquid phase（液相总传质单元高度）.

Appendix 11.12　Data recording and collation table of drying experiment
附录 11.12　干燥实验数据记录及整理表格

（1）Experimental data recording（实验数据记录）

Sample material（试样物料）_____；
Sample size（length × width × thickness）_____；
Moisture-free mass of sample（试样绝干质量）_____ g；
Initial mass of wet sample（开始时湿试样质量）_____ g；
Bracket mass（支架质量）_____ g；
Flowmeter indication value（流量计指示值）_____ mm；
Inlet control temperature of dryer（干燥器进口控温温度）_____ ℃；
Dry bulb temperature（干球温度）_____ ℃；
Wet bulb temperature（湿球温度）_____ ℃.

No. （序号）	Experimental data （实验数据）		Collation data （整理数据）		
	Mass of wet sample （湿试样质量），G_s/g	Time interval （时间间隔），$\Delta\tau$/s	Drying rate （干燥速率），U/[kg/(m²·s)]	Moisture content of wet sample （湿试样含水量），x	Mean moisture content （平均含水量）$\overline{X}_L = \dfrac{x_i + x_{i+1}}{2}$

(2) Experimental data collation (实验数据整理)

Mean air temperature(空气平均温度)t_m/℃	
Air wet bulb temperature(空气湿球温度)t_w/℃	
Volume flow of air(空气体积流量)V_t/(m³/h)	
Heating area of sample(物料受热面积)S/m²	
Mass flow of air(空气的质量流速)L/[kg/(m²·h)]	
Critical moisture content(临界含水量)X_c	
Equilibrium moisture content(平衡含水量)X^*	
Drying rate in constant-rate period(恒速段干燥流速)U_c/[kg/(m²·h)]	
Convective heat transfer coefficient in constant-rate period(恒速段对流传热系数)α/[W/(m²·h)]	
U_c estimated by port value(按进、出口值估计的U_c)/[kg/(m²·h)]	

Appendix 11.13 Data recording and collation table of the fluidized bed drying experiment
附录 11.13 流化床干燥实验数据记录及整理表格

(1) Determination of fluidization curve (流化曲线测定)

No.(序号)	Air temperature(空气温度)/℃	Air pressure(空气压力)/kPa	Pressure drop of orifice plate(孔板压降)/kPa	Bed pressure drop(床层压降)/kPa	Air flow(空气流速)/(m/s)

(2) Drying characteristic curve (干燥特性曲线) (x-τ, t_{bed}-τ, u-$\bar{x}$)

Specific surface area of mung beans $a = 1.23 \text{m}^2$ per/kg dry mung bean

No.(序号)	Sampling time(取样时间)/min	Bed layer temperature(床层温度),t_{bed}/℃	Mass of wet sample(湿物料重),G_s/g	Mass of dry sample(干物料重),G_c/g	Moisture ratio(含水率),x	Mean moisture ratio(平均含水率),$\bar{x}$	Drying rate(干燥速率)×10⁻⁴/[kg/(m²·s)]

Appendix 11.14 Data recording and collation table of extraction experiment
附录 11.14 萃取实验数据记录及整理表格

(1) Experimental data recording（实验数据记录）

① Experiment of changing the oil phase flow with fixed the rotational speed and water phase flow（固定转速和水相流量，改变油相流量实验）.

Rotational speed（转速）_____ r/min；Phase flow of water（水相流量）_____ L/h.

Flow rate of oil phase(油相流量)L_1/(L/h)	
Sampling volume of inlet oil phase(油相入口取样量)m_o/g	
Sampling volume of outlet oil phase(油相出口取样量)m_1/g	
Sampling volume of outlet water phase(水相出口取样量)m_2/g	
NaOH volume for titrating inlet oil phase sample(油相入口样品滴定用NaOH体积)V_o/mL	
NaOH volume for titrating outlet oil phase sample(油相出口样品滴定用NaOH体积)V_1/mL	
NaOH volume for titrating inlet water phase sample(水相入口样品滴定用NaOH体积)V_2/mL	

② Experiment of changing the rotational speed with fixed oil and water phase flow（固定油相流量和水相流量，改变转速实验）.

Phase flow of oil phase（油相流量）_____ L/h；

Phase flow of water phase（水相流量）_____ L/h.

Rotational speed(转速)r/(r/min)	
Sampling volume of inlet oil phase(油相入口取样量)m_o/g	
Sampling volume of outlet oil phase(油相出口取样量)m_1/g	
Sampling volume of outlet water phase(水相出口取样量)m_2/g	
NaOH volume for titrating inlet oil phase sample(油相入口样品滴定用NaOH体积)V_o/mL	
NaOH volume for titrating outlet oil phase sample(油相出口样品滴定用NaOH体积)V_1/mL	
NaOH volume for titrating inlet water phase sample(水相入口样品滴定用NaOH体积)V_2/mL	

(2) Experimental data collation（实验数据整理）

Experiment of changing the oil phase flow with fixed the rotational speed and water phase flow（固定转速和水相流量，改变油相流量实验）.

Mass fraction of benzoic acid in oil inlet solution（油入口溶液中苯甲酸的质量分率）$W_{t_0} \times 10^3$	
Mass fraction of benzoic acid in oil outlet solution（油出口溶液中苯甲酸的质量分率）$W_{t_1} \times 10^3$	
Mass fraction of benzoic acid in water inlet solution（水入口溶液中苯甲酸的质量分率）$W_{t_2} \times 10^3$	
Number of mass transfer units N_{OR} according to the raffinate（萃余相为基准的总传质单元数），N_{OR}	
Height of mass transfer units H_{OR} based on the raffinate（萃余相为基准的总传质单元高度），H_{OR}/m	
Total mass transfer coefficient based on raffinate（萃余相为基准的总传质系数），$K_x a/[kg/(m^3 \cdot h)]$	

Appendix 11.15 Data recording and collation table of pervaporation membrane experiment
附录11.15 渗透蒸发膜实验数据记录及整理表格

(1) Experimental data recording（实验数据记录）

Name of raw material liquid（原料液名称）_____；

Name of membrane material（膜材料名称）_____；

Membrane area（膜面积）_____ m^2.

No. （序号）	Feed temp. （进料温度），$t/℃$	Vacuum （真空度），P/MPa	Operating time （操作时间），$\Delta t/min$	Raw material liquid （原料液)		Permeate liquid （透过液)	
				Pre-test conc. （实验前浓度）C_{A1}	Test end conc. （实验结束浓度）C_{A2}	Concentration （浓度）r_A	Mass （质量），m/kg

(2) Experimental data collation（实验数据整理）

No. （序号）	Feed temp. （进料温度），$t/℃$	Vacuum （真空度），P/MPa	Operating time （操作时间），$\Delta t/min$	Concentration of feed liquid （原料液浓度）C_{A1}	Separation factor （分离因子）α	Pervaporation flux （渗透通量）$J/[kg/(m^2 \cdot min)]$

Appendix 11.16 Data recording and collation table of ultrafiltration membrane separation experiment
附录 11.16 超滤膜分离实验数据记录及整理表格

(1) Experimental data recording（实验数据记录）

Name of raw material liquid（原料液名称）_____

Name of membrane material（膜材料名称）_____

Membrane area（膜面积）_____ m^2

No. （序号）	Raw material liquid （原料液）		Permeate liquid （透过液）	
	Initial concentration （初始浓度）C_1	Volume （体积）V_1/L	Concentration （浓度）C_2	Volume （体积）V_2/L

(2) Experimental data collation（实验数据整理）

No. （序号）	Operating pressure difference （操作压差） ΔP/kPa	Experimental time （实验时间）， θ/h	Rejection rate （截留率） f	Separation factor （分离因子） α	Pervaporation flux （渗透通量）J /[kg/(m^2·min)]

References
参考文献

[1] 夏清,贾绍义.化工原理.天津:天津大学出版社,2012.

[2] 陈寅生.化工原理实验及仿真.上海:东华大学出版社,2008.

[3] Warren McCabe. Unit Operations of Chemical Engineering. New York:McGraw-Hill Education,2004.

[4] Unit Operations of Chemical Engineering 英文改编版.伍钦,钟理,夏清,熊丹柳,改编,北京:化学工业出版社,2019.

[5] 王春蓉.化工原理实验.北京:化学工业出版社,2018.

[6] 张金利,郭翠梨,胡瑞杰,范江洋.化工原理实验.天津:天津大学出版社,2016.

[7] 叶向群,单岩.化工原理实验及虚拟仿真(双语).北京:化学工业出版社,2017.

[8] 庞秀言,润明涛.化工基础实验.北京:化学工业出版社,2020.

[9] 伍钦,邹华生,高桂田.化工原理实验.广州:华南理工大学,2014.

[10] 吴晓艺.化工原理及工艺仿真实训.北京:化学工业出版社,2019.

[11] 贾广信.化工原理实验指导.北京:化学工业出版社,2019.

[12] 田维亮.化工原理实验及单元仿真.北京:化学工业出版社,2018.

[13] 代伟.化工原理实验及仿真(汉英对照).武汉:武汉大学出版社,2018.